Advance Praise for
Ghostriders 1968-1975

"*Ghostriders 1968-1975* provides the final word on the birth and Vietnam years of the AC-130 Spectre gunship. Bill Walter nails the legends and the everyday activities of the aircraft's developers and crewmen with uncanny accuracy. His accounts fulfill the highest standards of research. Reading took me back to the early '70s when I crewed with Spec and his facts filled a few gaps in my knowledge of the time. The book is a valuable addition to USAF history."

—Lt Col Hank Zeybel, USAF (Retired)

"The AC-130 Spectre gunship is without equal on the battlefield. It is the most precise and lethal form of combat airpower in the world.

"During the time that I had the privilege to serve as the AFSOC Commander, one generation of aging Spectres was preparing to leave the inventory as they were replaced with a newer generation of AC-130 gunships. The tremendous history and heroic actions of the crews who flew them and the maintainers who pushed them into the air deserved to be captured for posterity. We contracted an author to collect and publish this amazing history and were disappointed at the low quality of his output. That's when CMSgt Bill Walter stepped in to take over this important and historical project. His diligence to 'get it right' involved countless hours of interviews with both the crews and the soldiers on the ground, many of which would not be alive today, if not for the gunship. Every story herein is absolutely true, as told by those who were on scene. Multiple sources were confirmed for the details of every harrowing mission, and every dangerous engagement of the enemy.

"The orchestrated efforts of more than a dozen crewmembers to deliver lethal fires to destroy enemy targets or to protect the lives of American soldiers, sailors, airmen, and Marines on the ground is without equal. There are literally thousands of our combatants who owe their lives directly to the professionalism and skills of AC-130 Spectre gunships.

"My father is one of them. The very first AC-130 saved his life when the Spectre responded to a camp in Vietnam as it came under attack by two Viet Cong battalions. The heroism of the Spectre crew that night is the only reason my father survived. When I took command of the Special Operations wing at Hurlburt Field, thirty years later, my parents came to attend. The wing put all of the aircraft on display for the guests. Following the ceremony, I asked my dad if he would like to go see the MH-53J Pave Low helicopter that I flew as a special operations pilot. He said, 'I want to go see the gunship.' Like so many others who will never forget 'that night' that an AC-130 Spectre was overhead raining death on enemy forces to save their lives, he remembered those events for the rest of his life. Stories like that are legion. In the official Air Force records, the post-mission reports read something like, 'Responded to TIC (troops in contact). Fired 80 rounds 20 millimeter, 32 rounds of 40 mm, and 18 rounds of 105 mm.' But the people on the ground will tell an entirely different and desperate tale. This book captures those stories, as told by those who lived them.

"This will be the most popular book ever written regarding the aircraft and crews of our special operations forces. The culture and warrior ethos of the Spectre community is one of the strongest in Air Force history. You will be amazed and proud that such Americans serve to protect our freedoms."

—Lt Gen Donny Wurster, USAF (Ret.)

GHOSTRIDERS 1968-1975

WILLIAM WALTER

"MORS DE CAELIS" COMBAT HISTORY OF THE AC-130 SPECTRE GUNSHIP, VIETNAM, LAOS, CAMBODIA

KNOX PRESS

A KNOX PRESS BOOK
An Imprint of Permuted Press
ISBN: 978-1-63758-155-1
ISBN (eBook): 978-1-63758-156-8

Ghostriders 1968-1975:
"Mors De Caelis" Combat History of the AC-130 Spectre Gunship, Vietnam, Laos, Cambodia

Cover art by Cody Corcoran

Permuted Press, LLC
New York • Nashville
permutedpress.com

Published in the United States of America
1 2 3 4 5 6 7 8 9 10

This book is dedicated to the unsung professionals who operated, maintained, and supported the AC-130 gunship mission from 1968 through 1975.

The first production AC-130A Gunship to arrive for combat duty in Southeast Asia was tail number 54-1630, named *Mors De Caelis* (Latin translation Death From Heaven). Upon retirement in 1995, 54-1630 was inducted into the National Museum of the United States Air Force in Dayton, Ohio. It is currently on display under the name "Azrael, Angel of Death."

TABLE OF CONTENTS

THE AC-130A/E/H FLEET, 1968-1975[1]

Tail #	Name	Model	Status/Unit(s) of Assignment
53-3129	*The First Lady*	AC-130A	USAF, 16 SOS, 415 SOTS
54-1623	*Ghost Rider*	AC-130A	USAF, 16 SOS, 415 SOTS
54-1625	*The Warlord*	AC-130A	USAF, 16 SOS
54-1626	*Vulcan Express*[2]	Gunship II	USAF, ASD
54-1627	*Gomer Grinder*	AC-130A	USAF, 16 SOS, 415 SOTS
54-1628	*The Exterminator*	AC-130A	USAF, 16 SOS, 415 SOTS
54-1629	*The Arbitrator*	AC-130A	USAF, 16 SOS
54-1630	*Mors de Caelis*[3]	AC-130A	USAF, 16 SOS
55-0011	*Night Stalker*	AC-130A	USAF, 16 SOS
55-0014	*Jaws of Death*[4]	AC-130A	USAF, 16 SOS

1 Other names were known to have been used but names listed are the last known record.

2 Also called *Big Spooky* and *Super Spooky*

3 Later renamed *Azrael, Angel of Death*

4 Retired as *Jaws of Death*, but may have once been *Immortal*, *Cobra*, or *Widow Maker* (exact tail numbers for names unknown).

55-0029	*Midnight Express*[5]	AC-130A	USAF, 16 SOS
55-0040	*Orion the Hunter*	AC-130A	USAF, 16 SOS
55-0043	*Odin*[6]	AC-130A	USAF, 16 SOS
55-0044	*Prometheus*[7]	AC-130A	USAF, 16 SOS
55-0046	*Proud Warrior*[8]	AC-130A	USAF, 16 SOS
56-0469	*Grim Reaper*	AC-130A	USAF, 16 SOS
56-0490	*Thor*[9]	AC-130A	USAF, 16 SOS
56-0509	*Raid-Kills 'Em Dead*[10]	AC-130A	USAF, 16 SOS
69-6567	*Ghostrider*	AC-130H	USAF, 415 SOTS
69-6568	*Night Stalker*	AC-130H	USAF, 415 SOTS
69-6569	*Excalibur*[11]	AC-130H	USAF, 16 SOS
69-6570	*Bad Company*[12]	AC-130H	USAF, 16 SOS
69-6571	*The Destroyer*	AC-130E	USAF, 16 SOS
69-6572	*Grave Digger*	AC-130H	USAF, 16 SOS
69-6573	*Heavy Metal*	AC-130H	USAF, 16 SOS
69-6574	*Iron Maiden*	AC-130H	USAF, 16 SOS
69-6575	*Wicked Wanda*	AC-130H	USAF, 16 SOS
69-6576	*Hell Raiser*[13]	AC-130H	USAF, 16 SOS
69-6577	*Death Angel*	AC-130H	USAF, 16 SOS

---

5 Retired as *Midnight Express*, but may have once been *Immortal, Cobra*, or *Widow Maker* (exact tail numbers for names unknown).

6 Was call sign Spectre 11 when shot down.

7 Was call sign Spectre 13 when shot down.

8 Retired as *Proud Warrior* but may have once been *Immortal, Cobra*, or *Widow Maker* (exact tail numbers for names unknown).

9 Was prototype "Surprise Package" aircraft; also known as *Super Chicken*. Was call sign Spectre 17 when shot down.

10 Retired as *Ultimate End*

11 Was also *Fatal Attraction*

12 Was also *Widow Maker, The Hussy*

13 Was also *Predator*

FOREWORD

The United States Air Force (USAF) fixed-wing, side-firing gunship program began in the early 1960s during the Vietnam War with the development, fielding, and operational use of the AC-47D *Spooky* gunship. Contrary to typical aircraft development, the gunship program was not devised, controlled, or marketed by the commercial aviation industry. Instead, the Air Force Gunship Program Office and Air Force Systems Command (AFSC) took matters into their own hands to rapidly develop and field a desperately needed combat capability. Without a doubt, the side-firing fixed-wing gunship program would have failed if not for the leadership and savvy of the USAF Gunship Program Office, Air Force Systems Command (AFSC), and Aeronautical Systems Division (ASD) engineers. Furthermore, program success was a direct result of the heroic and unsung efforts of *Spectre* aircrews who flew thousands of dangerous missions over Vietnam, Laos, and Cambodia.

Overall, the accomplishments of the combined team of engineers, maintainers, and aircrews were especially commendable since they inherited some of the oldest, broken-down aircraft in the USAF inventory. Nevertheless, gunship crews developed tactics, techniques, and procedures for close air support, interdiction, command and control, and even search and rescue. By the end of the war, the AC-130 had become the most revered attack aircraft of the Vietnam war, but the classification of their deeds prevented public release. Now, over fifty years after the first AC-130 Gunship began flying combat missions over Vietnam, the declassified, factual and comprehensive story of *Spectre* can be told.

Original members of the ASD Gunship II development team at Wright Patterson AFB, Ohio, 1967, in front of the prototype aircraft. L–R, Technical Sergeant Bob Davis, Major Jim Krause, Major Jim Wolverton, Major Ron Terry, Major Harris, Master Sergeant Farris Heins, Technical Sergeant Paul Bunch. (USAF Photo)

PREFACE

Beginning in the early 1960s, the number of U.S. ground forces steadily increased in Vietnam. Numerous U.S. Army Special Forces camps were situated in South Vietnam to protect the citizens of South Vietnam from attack by North Vietnamese insurgents. These small SF camps protected hamlets and villages came under constant attack by communist Viet Cong guerrillas. Though most attacks were conducted by small teams, large attacks of one thousand or more had occurred whereas SF camps were overrun and defenders killed. It was evident U.S. Army Special Forces desperately needed persistent, lethal Close Air Support (CAS) to defend against Viet Cong human wave attacks, especially at night.

Unfortunately, a capabilities gap existed since the majority of USAF fighter aircraft were designed to fight a third world war scenario against the Soviet Union. The USAF quickly discovered fighters capable of supersonic speed with short mission duration (because of fuel burn rates) proved difficult to employ in a slow-speed, long time-on-target role. Once USAF leadership (Air Staff) recognized the capabilities gap, WWII era propeller-driven fighter-bomber aircraft were pressed into service in Vietnam. These aircraft were more appropriate for the CAS role since they were slow and could remain over target for long periods. Though these aircraft were primarily day only, the use of illumination flares provided a rudimentary night capability.

Recognizing the issues at hand, Aeronautical Systems Division (ASD) Engineers at Wright Patterson Air Force Base (AFB), Ohio brainstormed ideas to quickly improve CAS capabilities in Vietnam. Acknowledging the fundamental capabilities of an effective CAS aircraft are persistence, lethality, and survivability, the challenge to find a solution via conventional means was difficult. Recognizing the dilemma, engineers believed the "pylon turn" was a logical concept. This tactic required the pilot to fly a circular orbit over a fixed point on the ground. In effect, the pilot would fly a constant left banked turn for extended periods to ensure constant, uninterrupted sight of the target or ground force. Firepower would be provided by multiple guns and a large amount of ammunition and fuel to remain on station for long periods of time. With all engineering tenets and constraints considered, converting a cargo aircraft into a gunship was considered the best option. Though ASD Engineers initially met resistance at Air Staff, by 1964, they received a small budget to proceed. Initial experiments were conducted on a C-131 transport, first using a camera, then a gun. The experiment demonstrated promise to move the experiment into a project phase.

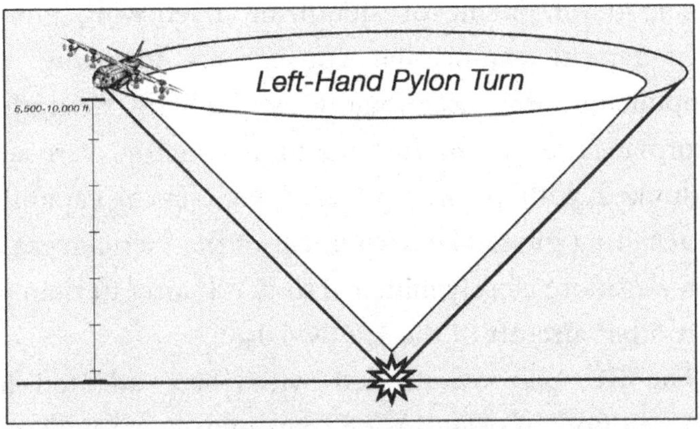

The basic pylon turn concept requires the pilot to maintain a constant left-hand bank, encircling a target. Once aligned and at the proper airspeed and bank angle, the ballistic targeting solution remains relatively constant, enabling constant fire on the target for as long as needed. The firing solution, referred to as on-nominal geometry was the critical aspect of gunship effectiveness proven by all models of gunships during the Vietnam War. (Illustration by Dennis Bivens and Sarah Walter)

Once ASD engineers proved the concept viable, the WWII era C-47D was chosen for conversion since it was available and already in use supporting unconventional warfare in Vietnam.

> *The Air Force carefully weighed the combat advantages and disadvantages of this C-47 with laterally-firing guns. The aircraft was available and so were the crews to fly it. The plane could carry a large load of ammunition and flares and could be used for cargo, troop, and reconnaissance missions. It possessed two engine safety, long alert capability, lengthy time-over-target, and the capability to loiter for flare dropping.* Development and Employment of fixed wing Gunships, Jack S. Ballard P-11.

The development of side-firing fixed-wing gunships defied typical engineering conventions in many ways. Components were intentionally overbuilt for strength and performance. Crew comfort and human factors were largely overlooked with priority focused on tactical capabilities. The resulting physical environment of the Vietnam era gunships was more closely matched to WWII aircraft than modern combat aircraft of the 1960s–70s.

The AC-47D was armed with three adapted SUU-11/A gun pods (GAU-2/A 7.62mm mini-guns) with a normal combat ammunition load exceeding twenty thousand rounds of tracer/ball ammunition. The aircraft also carried a quantity of MK-24 illumination flares for nighttime target acquisition.

The AC-47D gunship made its Vietnam combat debut in December of 1964. Air and ground crews of the 1st Air Commando Wing developed combat tactics via experimentation. Typical operational altitude was three thousand feet, which provided adequate 7.62mm lethality, but also placed the gunship in the effective range of enemy small arms, anti-aircraft artillery (AAA), and even mortar fire. Nevertheless, the AC-47D gunship was successful, especially when employed in a CAS/Troops In Contact (TIC) role against personnel targets. In 1966, North Vietnamese vehicle traffic on the Ho Chi Minh Trail increased, and AC-47D gunships were sent into Laos to hunt trucks. Unfortunately, *Spooky's* low operational altitude and lack of heavy firepower resulted in minimal truck damage and loss of several aircraft and crews. It was obvious the Ho Chi Minh Trail was not a viable AC-47D mission.

In 1967, The ASD Gunship Program Office set forth an effort to upgrade proven AC-47D capabilities with a larger and more powerful aircraft, and the C-130 was favored for conversion. On 2 February 1967, C-130A serial number 54-1626 assigned to Edwards Air Force Base as a test asset was flown to Wright Patterson Air Force Base for modification. Known as "Sick Two Six", it had been in three major accidents since it was built, but was still airworthy and considered a good starting point.

Development of the prototype C-130 gunship began on 1 April 1967 at Wright-Patterson AFB, Ohio, under the name *Project Gunboat.*[14] In concept, it was designed to assume the AC-47D mission role but at higher altitudes, with a greater payload and better lethality. Like its AC-47D predecessor, the development of the prototype AC-130A gunship focused more on capability and less on human factors. To facilitate rapid development, existing doors and hatches were used to mount equipment to the maximum extent possible. The conventional C-130 crew entrance door was replaced with a half-door closeout panel to enable mounting of a night observation device (NOD) visual sensor. A state-of-the-art infrared (IR) detector set was installed on the floor which peered through a hole cut in the fuselage forward of the left wheel well. The IR Sensor Operator sat in an enclosed booth inside the cargo compartment. The right-side escape hatch was removed and the rear cargo door was fixed in the open position to provide scanners an unobstructed view

14 The project name was later changed to Gunship II.

to the left, behind, and below the aircraft. To enable effective scanning for AAA threats, an Illuminator Operator (IO) hung the top half of his body over the edge of the open ramp with only a restraining cable attached to his parachute harness. The Illuminator Operator also dispensed flares from a LAU-74 launcher and operated a large forty kilowatt (KW) searchlight, both of which were mounted on the ramp. The searchlight could be set for either visible or covert (IR/UV) illumination modes but was cumbersome and difficult to use.[15]

The initial configuration was named *Vulcan Express* which included eight guns (four 20mm and four 7.62mm) bristling from the left side of the fuselage. A Head-Up Display (HUD) gunsight was mounted in the left window of the cockpit to enable the pilot to aim the guns. The HUD and gun mounts were adjustable for depression and lag and could be set for 2,500; 3,500; 4,500; and 5,500 feet altitudes. All guns were driven by electric motors powered by large rechargeable batteries. Each of the four M61 *Vulcan* 20mm guns fired linked ammunition at a cyclic rate of 2,500 rounds per minute from a 1,500-round capacity can. The usual 20mm ammunition load was 6,000 rounds (four full feed systems), but additional cans of ammunition could be carried onboard for in-flight reloading. Two types of ammunition were available for use: High Explosive Incendiary (HEI) and Armor Piercing Incendiary (API) with HEI being the predominant combat round. As the guns fired, spent cas-

15 The illuminator was comprised of two 20KW Xenon lamps mounted in parallel.

ings and links were ejected into a plywood walled bin on the floor. Once ejected, a gunner cleared brass and links away from the underside of the gun with a common grain shovel, and secured casings and links into canvas bags.

Four 7.62mm guns were mounted on MXU-470 modules located both forward and aft of the left wing. The modules were specially designed for AC-47 and AC-130 use via modification of inventoried SUU-11/A gun pods. Each module was equipped with a GAU-2/A gun and a link-less helical feed system with a capacity of two thousand rounds of Ball/Tracer ammunition. Linked cartridges were loaded into the system's helix via a de-linking feeder. The typical combat load ranged between fifteen thousand to sixteen thousand rounds, but upwards of twenty-four thousand rounds could be carried if required.

During a typical engagement, only one caliber/pair of guns would be fired at a time while the other pair remained ready in the event of a malfunction or depletion of ammunition. Though all guns could be fired simultaneously in a "full broadside" attack, that level of effort was not advised since the guns would overheat, increasing ground force vulnerability while guns either cooled down or were being reloaded. Another issue related to "full broadside" firing was the fact that smoke generated quickly filled the fuselage. Unfortunately, airflow resulting from the open cargo door caused smoke to blow forward into the cockpit, hindering the vision of the pilots, flight engineers, and navigators. To mitigate this phenomenon, ventilation ports were installed in the fuselage near each gun station to force air into the fuselage and redirect smoke to the rear and out of the cargo door.

Since *Gunship II* was being built for higher altitude operations, engineers deemed simple ballistic calculations to be problematic. In response, ASD engineers designed and built a basic fire control computer for *Gunship II*.

> "The Air Force in-house developed fire-control system, which receives inputs from the NOD, FLIR and SLR, provides aiming and steering information to the pilot. The fire-control computer integrates the inputs from any one of the three sensors to establish a line of sight (LOS) to a designated point. It compares this LOS with the corrected gun-line (CGL), the corrected gun line being the path through space which will be taken by a projectile fired at that instant, and provides position and altitude guidance information visually to the pilot through a gunsight and ILS indicator. The pilot flies the aircraft to enter a search or attack orbit accordingly. Once in the attack orbit, the pilot maneuvers the aircraft to place the fixed pipper in his gunsight (CGL) on the target reticle in his gunsight (LOS) to fire the weapon. The fire-control computer also corrects for aerodynamic wind, true airspeed and altitude."
> 16 SOS Unit Record, 1 January 1969.

On 15 September 1967, the *Gunship II* Task Force deployed with aircraft 54-1626, now named *Big Spooky* to Vietnam for a sixty to ninety day combat evaluation.

> "Headquarters USAF directed that the Tactical Air Command deploy and accomplish combat evaluation of the Gunship II weapon system in Southeast Asia (SEA). One C-130 Aircraft was used in the evaluation, which began during the latter part of September, 1967. The C-130A Gunship, later designated the AC-130, was developed as an in-house project at Wright-Patterson AFB, Ohio. Operational test and evaluation was conducted at Eglin AFB, FL in June, August, and September, 1967. The Gunship II Task Force was deployed to SEA on 15 September 1967. Operations during September – December were conducted in South Vietnam." Unit History Report, 16 SOS 1 Jul-30 Dec, 1968. (U)

Initial *Gunship II* missions and tactics were essentially a carbon copy of AC-47 operations with exception of a few thousand feet additional altitude capability and larger 20mm guns. *Gunship II* was tested in three phases. The first was devoted to close air support missions from airborne alert in the Delta region–IV Corps Tactical Zone (CTZ) around Binh Thuy. The second phase tried the *Gunship II* weapon system against enemy lines of communication in

Tiger Hound. The third phase involved armed reconnaissance and ground support missions in the highlands of II Corps CTZ.[16]

Gunship II test results also proved to be an effective truck killer on the Ho Chi Minh Trail, sighting ninety-four and destroying thirty-eight. General William Moore, Air Force Deputy Chief of Staff Research and Development commented, "the C-130 *Gunship II* test bed aircraft had unprecedented success in identifying and destroying enemy lines of communication in both South Vietnam and Laos," and that it "far exceeded fighter-type kill ratios on enemy trucks and other equipment."

When combat trials concluded in December of 1967, aircraft 54-1626 returned to Wright-Patterson AFB, OH. When results were briefed to General John D. Ryan, Commander In Chief Pacific Air Forces (CINPACAF), he was so impressed that he pushed for full-scale development of a fleet of AC-130s. With high-level endorsement by General Ryan, ASD pursued funding for the development of thirty AC-130s to augment and ultimately replace the AC-47D *Spooky* gunship.

When Air Staff was briefed on *Gunship II*, combat trials, the concept of a costly side-firing gunship program was unwelcome. Nevertheless, General Ryan was adamant that Gunship II was the breakthrough combat capability desperately needed to fight the war in SEA. Finally, in February of 1968, Secretary of the Air Force, Dr. Harold Brown, directed the gunship conversion of seven JC-130A aircraft stationed

16 Development and Employment of Fixed-Wing Gunships, by Jack S. Ballard P-86.

at Patrick Air Force Base, Florida. These aircraft were some of the oldest, most broken-down C-130s in the inventory which were slated for retirement.[17] The remaining requirement to replace the AC-47D would be accomplished by converting two equally old models of the C-119 into gunships.

General Ryan however, did not favor a mixed fleet as he indicated in his letter to SECAF.

> "Recent highly successful combat evaluation Gunship II favors AC-130 as logical replacement for AC-47. AC-130 possesses needed capabilities as follows:
>
> Speed (rapid reaction, area coverage, minimum exposure).
>
> Sensors (locate enemy and friendly positions, deliver accurate firepower)
>
> Increased payload (essential to carry increased firepower, sensors, armor)
>
> Further advantages of C-130 are superior performance/flexibility, worldwide maintenance/supply support, contemporary navigation systems, established pilot training, schools and post-hostility airframe re-conversion potential....Gunship II C-130s should not be considered at the expense of current and projected airlift assets. New pro-

17 Aircraft 53-3129 had the distinction of being the first production C-130 aircraft and was also one of the first four production AC-130s.

> duction C-130 aircraft appears to be warranted in view of recent mortar attacks on forward installations. Requirement for 32 AC-130 gunship force...considered urgent as is provides the most effective reaction capability against attack on installations. Recommend reconsiderations C-130 as follow-on gunship for AC-47 on a one-for-one basis." Unit History Report, 16 SOS 1 Jul-30 Dec, 1968.

Members of the Gunship II Task Force in Vietnam. (USAF photo)

In February 1968, while ASD engineers began their effort to convert the first four JC-130A aircraft into AC-130 gunships, an initial cadre of volunteers and members of the ASD task force flew the *Gunship II* prototype back to SEA. Their responsibility was to lay the groundwork for establishment of an operational squadron of AC-130 gunships.

The official designation of the first seven production AC-130 gunships was *Plain Jane*, a fitting title since the basic program was fundamental and "plain" with spikes of high technology. Other than basic C-130 knowledge, aircrews and maintainers were not familiar with numerous commercial avionics systems installed, therefore, Contractor Logistic Support (CLS) specialists were assigned to the program. Problems arose quickly since there was no AC-130 training program, technical orders, equipment checklists, or spare parts for most of the high-technology equipment installed on the gunship. Since these issues were substantial "show stopping" problems, ASD had to resolve or mitigate many problems in a short period of time.

Selecting the right AC-130 aircrew and maintainers was absolutely critical to the future success of combat employment. These men would have to "think on their feet" not only to operate and maintain the unique systems but would also develop initial procedures and tactics, the hard way, during combat operations in SEA. Most of the original non-ASD cadre were volunteers from the JC-130 unit at Patrick AFB that aircraft were being pulled from to build AC-130s. At about the same time, Ling-Temco-Vought (LTV) in Greenville, Texas, won a contract to produce seven AC-130s: two for CONUS training and five for combat duty in SEA. Simultaneously, a formal AC-130 gunship training program was enacted at the 4413th Combat Crew Training Squadron at Lockbourne AFB, Ohio. A lot of faith was riding on the first seven aircraft, and there was a great deal of technical and physical risk for all involved in the program. Fortunately, the ASD engineers were not deterred by those who viewed their ad-hoc project foolhardy and misdirected. While the first seven AC-130s were being built, arguments centered on the number and type of gunships needed were a hotly debated topic throughout the USAF and PACAF chain of command. By the end of the war in Vietnam, the AC-130 gunship program had proven its worth as a truck killer, accounting for over ten thousand vehicles destroyed. The People's Army of Vietnam, specifically *Group 559* who transported supplies from North Vietnam through Laos and Cambodia, cited the AC-130 as the most feared aircraft of the entire war.

The AC-130 program was successful because it provided the right capability at the right place at the right time. Conventional thinkers may still believe the fixed-wing gunship concept blurs roles and missions of fighter, bomber, and cargo aircraft and is vulnerable to enemy threat weapons. In that regard, the AC-130 is a liability, but when employed within its operational parameters, the AC-130 provides a capability that is second to none. In the end, as with all combat aircraft, it's not "what it is" that's important, it's "what it does."

AC-130A "GUNSHIP II" AND "PLAIN JANE"

AC-130A 54-1629 within weeks of arrival at Ubon. Note the tail code "EA" signifies assignment to 14th Air Commando Wing, Nha Trang, Vietnam. (USAF Photo)

AC-130A Gunship II and Plain Jane	Crew
Bell Optical Gun Sight	Pilot/Aircraft Commander (AC)
20mm M61 gun (x4)	Copilot (CP)
7.62mm MXU-470 gun (x4)	Flight Engineer (FE)
AN/AVG-2 Night Observation Device (NOD)	Navigator (NAV)
AN/APQ 133 Side Looking, Beacon Tracking Radar	Sensor Operator (IR and Radar)
AN/APQ-136 Forward Looking Radar	Sensor Operator (NOD)
AWG-13 Fire-control System computer	Illuminator Operator/ Loadmaster (IO)

AN/AVQ-8 40KW Steerable Illuminator	Aerial Gunner (AG) (x3)
MK-24 Flare Launcher	Combat Photographer
AAD-4 Forward Looking Infrared (FLIR) Set	
AN/APN-81 Doppler Navigation Radar	
Inert polyurethane foam filled fuel tanks	
ID-48 Steering Indicator	
Ceramic Armor Plating	
FM Radio transceiver	
TRIM-7 Active ECM	

GUNSHIP II AND AC-130A PLAIN JANE BASIC CREW DUTIES

Aircraft Commander (AC): Has the overall responsibility for coordinating the mission. Flying from the left seat of the cockpit, the AC attains firing geometry, balances the gunsight, controls aircraft bank, fires the guns, and has the ultimate responsibility for putting firepower on target. All original cadre pilots were qualified Aircraft Commanders.

Second Pilot: Assumes the duties and responsibilities of Aircraft Commander as necessary. The second pilot also controls aircraft speed during the firing mode. The second pilot performs normal copilot duty with special emphasis on outside traffic and AAA activity.

Third pilot (P): Responsible for setting the fire-control computer and was the Fire Direction Officer (FDO).[18] The Third Pilot supported other crew positions, especially the NOD operator who was required to stand in an open doorway for the duration of the mission (*Gunship II*) however, *Plain Jane* models were equipped with a rudimentary seat. The third pilot position was used extensively on *Gunship II* but eventually faded away due to manpower requirement changes and technology improvements on production gunships.

Navigator (NAV a.k.a. Table Nav): Has the responsibility for en route navigation, aircraft positioning in the target area, and aircraft traffic separation. The NAV confirms

---

18 The Third Pilot/Fire Direction Officer position was used on the early "plain Jane" configuration, but the fire-control computer was problematic and rarely worked as designed.

target location, separation from friendly forces, and clears the pilot to fire. Navigators routinely filled positions of the NOD and IR operator also.

Flight Engineer (FE): Operates systems controls and observes engine instruments, systems indicators, and control devices and reports abnormal conditions to the pilot. The FE also monitors and controls the gun control panel as directed by the pilot and coordinates between the pilot and gunners.

Master Armorer/Aerial Gunner (AG):[19] Gunners man three positions in the aircraft. Forward and aft gunners operate, repair malfunctions, reload and remove spent casings from all gun systems as required. The Master Armorer/Lead Gunner is responsible for coordinating and leading gun crew efforts. One gunner performs duties of right scanner to detect AAA fire and airborne traffic and direct defensive maneuvers.

Infrared Sensor Operator (IR): Was usually a navigator. Has the responsibility for operating and monitoring the infrared system to search, locate and track targets. When selected as the primary sensor operator, the IR provides guidance to the fire-control system. The IR coordinates with other sensor operators and the Fire Direction Officer (FDO) in the validation of targets and the assessment of damage inflicted on the target.

---

19 Though duties remained the same, the title of this position changed over the years. In 1974, gunners were referred to as "Weapons Mechanics" since they were drawn from the 462X0 "Weapons Mechanic" career field. An assignment to AC-130s was special duty in the 462X0 career field, but they were still controlled as Weapons Mechanics.

Night Observation Device (NOD) Operator: Was usually a navigator. Has the responsibility to operate the NOD to search, locate and track targets. When selected as the primary sensor operator, the NOD provides guidance to the fire-control system. The NOD coordinates with other sensor operators and the FDO in the validation of targets and the assessment of damage inflicted on the target.

Illuminator Operator (IO):[20] Operates 40KW Searchlight, dispenses flares and markers. Performs aft scanner duties from the open aft door advising the crew of ground fire and other airborne traffic and directs defensive maneuvers and performs the duties of loadmaster as required.

Aerial Photographer: Shoots 16mm motion picture film for bomb damage assessment (BDA) intelligence.

The Illuminator Operator hung the upper half of his body over the end of the open ramp when over hostile territory. This seemingly awkward position enabled a very wide field to detect hostile fire and direct the pilot to perform evasive maneuvers. (USAF Photo)

20 By 1970, the 40KW Illuminator was not favored or used by combat crews. It was replaced by the LAU-74 illumination flare launcher, but the title "Illuminator Operator" remained in place until the mid-1990s. Original IOs were Aircraft Crew Chiefs assigned to flying positions via special duty. Not until the 1990s were IOs cross trained to the loadmaster career field.

SURPRISE PACKAGE AND PAVE PRONTO, INTRODUCED 1970-1971

The one and only Surprise Package aircraft was 56-0490 known as "Thor." (USAF Photo)

Gunners loading 40mm gun on 56-0490 in February 1970. (Photo Dennis Bivens)

AC-130A Surprise Package & Pave Pronto	Crew
Optical Gun Sight	Pilot/Aircraft Commander (AC)
20mm M61 gun (x2)	Copilot (CP)
7.62mm MXU-470 gun (x2)	Fire-control Officer (FCO)
40mm M1/M2A1 (mod) gun (x2)[i]	Flight Engineer (FE)
AN/APQ-133 Side-Looking Beacon Tracking radar	Navigator (NAV)
APQ-136 Forward-Looking radar	Navigator/Sensor Operator (IR and Radar)
AN/ARN-92 LORAN Navigation Set	Navigator/Sensor Operator (LLLTV)
Digital Fire-Control Computer[ii]	Navigator/Electronic Warfare Officer (EWO)
LTN-51 Inertial navigation & targeting system	Illuminator Operator (IO)
UVR 700-AR Low-Light-Level-Television sensor[iii]	Aerial Gunner (x5) (AG)
2 KW Steerable Illuminator	
LAU-74 Flare dispenser	
AN/AAD-6/7 Forward Looking Infrared set	

i. The prototype 40mm guns installed were US Army/Navy M1 guns. Aircraft 56-0490 was the only AC-130 to mount M1 guns.
ii. Pave Pronto used AKY-9 Fire-control Syst*em Computer*
iii. Pave Pronto used AX*Q-10 LLLTV*

AN/ASD-5 Black Crow sensor	
Helmet sight[iv]	
ALQ-87 Electronic Warfare pods	
Ceramic armor plating	
Inert polyurethane foam-filled fuel tanks	
Doppler Navigation radar	
ID-48 Steering indicator	
AGMTIP Moving-target indicator	
WR-201 Video Bomb-Damage-Assessment-recorder	
KY-20A Laser Target Designator	
TRIM-7 active ECM	
AN/APR-25/26 RHAW	
Laser 8451-840000-1	

iv. Tested on Coronet Surprise only, *not installed on Pave Pronto.*

PAVE SPECTRE AND PAVE AEGIS, INTRODUCED 1971-72

Pave Spectre AC-130E aircraft 69-6571 modified with the Pave Aegis gun system and electronic countermeasure pods. (USAF photo)

AC-130E Pave Spectre	Crew
Optical Gun Sight	Pilot/Aircraft Commander (AC)
20mm M61 gun (x2)	Copilot (CP)
7.62mm MXU-470 gun (x2)	Fire-control Officer (FCO)
40mm M2A1 (mod) gun (x2)	Flight Engineer (FE)
105mm M137 gun (Pave Aegis mod only)	Navigator (NAV)
AN/APQ-150 Beacon Tracking Radar	Navigator/Sensor Operator (IR and Radar)

APN-59 Forward Looking Radar	Navigator/Sensor Operator (LLLTV)
ASN-90 -Inertial Navigation System	Navigator/Electronic Warfare Officer (EWO)
ASN-91 Fire-control Computer	Illuminator Operator (IO)
AXQ-10 LLLTV Sensor	Aerial Gunner (AG) (x5)
2 KW Steerable Illuminator	
LAU-74 Flare Dispenser	
AN/AAD-7 Forward Looking Infrared Set	
AN/ASD-5 Black Crow Sensor	
AN/ALQ-87 Electronic countermeasure pods	
Ceramic Armor Plating	
Inert polyurethane filled fuel tanks	
Nav System S431-1	
Direction-Finding Homing avionics	
FM Radio transceiver	
Pave Mace	
TRIM- 7 and AN/APR-25/26 Warning Receiver	
Laser 8451-84000-1	

AC-130A/E (SURPRISE PACKAGE/PAVE PRONTO, PAVE SPECTRE) BASIC CREW DUTIES[21]

Aircraft Commander (AC): Has the overall responsibility for coordinating the mission. Flying from the left seat of the cockpit, the AC attains firing geometry, balances the gunsight, controls aircraft bank, fires the guns, and has the ultimate responsibility for putting munitions on target.

Copilot (CP): Assumes the duties and responsibilities of pilot as necessary. The copilot controls aircraft speed, verbally directs aircrew checklists, handles outside communications with escorts, and controls the aircraft altitude should the autopilot fail.

Navigator (NAV): Has the responsibility for en route navigation, aircraft positioning in the target area and aircraft traffic separation. The NAV confirms target location and pre-established no-fire areas.

Fire-control Officer (FCO) (Surprise Package and Pave Spectre only): Monitors the fire-control display and coordinates crew effort in detecting, validating, and engaging targets. The FCO is also responsible for computing and compensating for ballistic wind and gun/sensor alignment errors.[22]

---

21 Tactical Air Command Formal Flying and Related Training Publication, student text, USAF Operational Training Course AC-130H Weapons Mechanics, Department of the Air Force, November, 1974.

22 The Fire Direction Officer position was on the early "Plain Jane" configuration. The name of the position was changed to Fire-control Officer on all subsequent variants.

Flight Engineer (FE): Operates systems controls and observes engine instruments, systems indicators, and control devices and reports abnormal conditions to the pilot. The FE also monitors and controls the gun control panel as directed by the pilot and coordinates between the pilot and gunners.

Lead Gunner/Aerial Gunner (AG):[23] Gunners man three positions in the aircraft. Forward and aft gunners operate, repair malfunctions, reload and remove spent casings from all gun systems as required. The Lead Gunner is responsible for coordinating and leading gun crew efforts. One gunner performs duties as right scanner to alert the crew of AAA fire and airborne traffic. The right scanner directs defensive maneuvers as required.

Infrared Sensor Operator (IR): Has the responsibility for operating and monitoring the infrared system to search, locate and track targets. When selected as the primary sensor operator, the IR provides guidance to the fire-control system. The IR coordinates with other sensor operators and the FDO/FCO in the validation of targets and the assessment of damage inflicted on the target.

Low-light-level-television Sensor Operator (TV): Coordinates with other sensor operators, navigator, and FCO, acquiring, tracking and maintaining proper target orientation. The TV also assists in the assessment of damage inflicted on the target.

23 Though duties remained the same, the title of this position changed over the years. In 1974, gunners were referred to as "Weapons Mechanics" since they were drawn from the 462X0 "Weapons Mechanic" career field. An assignment to AC-130s was special duty in the 462X0 career field, but they were still controlled as Weapons Mechanics.

Electronic Warfare Officer (EWO): Operates the Black Crow sensor to locate hostile targets, and beacon tracking radar to track friendly forces. The EWO operates all ECM equipment, advises the crew of threats and directs appropriate countermeasures.

Illuminator Operator (IO):[24] Performs aft scanner duties advising the crew of ground fire and other airborne traffic and directs defensive maneuvers and performs the duties of loadmaster as required.

A typical AC-130 crew circa 1969. The aircraft in the background is aircraft 54-1630 *Mors De Caelis*, which has three Purple Hearts painted on the fuselage to acknowledge battle damage on three separate occasions. 20mm ammunition silhouettes were painted on the fuselage, one round for every twenty missions. (USAF Photo)

---

24 By 1973, the 40KW Illuminator had been removed from all AC-130s and the LAU-74 illumination flare launcher was in limited use, but the title "Illuminator Operator" remained in place until the mid-1990s. Original IOs were Aircraft Crew Chiefs assigned to flying positions via special duty. Not until the 1990s were IOs cross trained to the loadmaster career field.

Crewmen were equipped with survival equipment including:

1. Survival Vest
2. Basic first Aid Kit
3. Compass and map of working area
4. Survival radio w/extra battery (sometimes two radios)
5. Tourniquet
6. MK-13 signal flares
7. Gyrojet launcher and flares
8. Signal Mirror
9. Water purification tablets
10. Model 15 or Model 10 revolver and 48 rounds of .38 ammunition
11. Parachute
12. Ballistic helmet and/or communications headset
13. Survival knife

SITUATION REPORT: VIETNAM, LAOS, AND CAMBODIA, 1968–1973

When *Gunship II* arrived at Ubon Royal Thai Air ForceBase in February of 1968, the civil war between South and North Vietnam was an ongoing major conflict. On 30 January, North Vietnam launched the *Tet Offensive*: their largest military campaign to date into South Vietnam, striking more than one hundred towns and villages. While this operation ensued, nearly one-half million US and allied personnel were in Vietnam and Thailand supporting the Army of the Republic of Vietnam (ARVN) who were conducting direct combat operations against the North Vietnamese. At the same time, China and the Soviet Union supplied weapons, equipment, and advisors to North Vietnam's People's Army of Vietnam (PAVN). Neighboring countries Laos and Cambodia had problems of their own as ethnic/political groups waged war against each other. Dissimilar factions in Laos and Cambodia were split on their support for North and South Vietnam, which only made matters worse.

The Royal Laotian government supported South Vietnam while Pathet Lao communists supported North Vietnam. North and South Vietnam were separated by a narrow Demilitarized Zone (DMZ) that created a significant strategic chokepoint for North Vietnamese troops crossing into South Vietnam. Realizing this insurmountable limitation, North Vietnam capitalized on the turmoil in Laos to create the Truong Son Road, a series of roads and trails known to the US as the Ho Chi Minh Trail.

The title "Ho Chi Minh Trail" was a misnomer since it was not a single trail, but a complex set of roads, trails, and hidden enclaves throughout the eastern side of Laos and into Cambodia. The entire trail network was organized and operated by thousands of soldiers augmented by North Vietnamese men and women laborers assigned to PAVN *Group 559*. Since labor required to operate the logistic system of the trail was tremendous, the PAVN sometimes captured tribal Laotians, indenturing them into forced labor services. *Group 559* operated multiple truck transportation regiments, and a petroleum pipeline that ran the entire length of the trail. Anti-Aircraft Artillery (AAA) and engineer regiments supported *Group 559* operations while the 968th Infantry Division provided security. Multiple Binh Trạm logistics support areas were situated along the entire length of the trial. Binh Trams were typically spaced between twenty to thirty kilometers apart to organize and direct truck relays from one to the next, minimizing exposure to attack.

Each Binh Tram was assigned a commander who directed an organized, controlled movement of truck traffic via field phone contact with the next one in line. PAVN personnel were assigned to a single Binh Tram for extended periods. Since drivers took the same routes nightly, they knew every twist, turn, and pothole in their leg of the trail. During a routine mission, drivers sped from their home Binh Tram to the next, unloaded, then returned to their home location as quick as possible. Supplies and fresh troops were trucked southward while wounded soldiers were trucked northward. Some Binh Trams were located near high terrain, some were dug into the side of mountains or located in natural caves, while others were simply concealed under triple canopy jungle. A few locations were constructed with underground tunnels containing operations, dining, medical and rest facilities; however, some were relatively primitive with only food, water, and fuel. In all locations, PAVN combat engineers dug fox holes for personal shelter and bomb fragment protection and built camouflaged log and dirt bunkers for their trucks. Most Binh Tram were well protected with AAA guns, but those located in triple canopy jungle were less defended.

The Ho Chi Minh trail logistics network was vitally important to North Vietnam to insert combat forces, supplies, and equipment into South Vietnam. Since the early 1960s, the Ho Chi Minh Trail had been systematically bombed, mined, sprayed with defoliant agents, and targeted with Commando raids during operations in *Barrel Roll*, *Steel Tiger*, and *Tiger Hound* areas of Laos. These efforts ran parallel to the strategic bombing of North Vietnam, and by the beginning of 1968 more than 100,000 fighter and bomber missions had been flown into Laos. To date, more than 130 US aircraft and helicopters had been shot down while flying missions over the trail. In late 1968, the Johnson administration halted bombing of North Vietnam, shifting its focus on cutting off North Vietnam's supply network to the south. Thus, in late 1968, the first full-scale interdiction effort of the Ho Chi Minh Trail began under the code name *Operation Commando Hunt*. This operation centered on locating and destroying supply depots, transload areas, trucks, supplies, and troops filtering down the Ho Chi Minh Trail towards South Vietnam.

In the years prior to the beginning of *Commando Hunt*, large PAVN truck convoys traveled at night to avoid being attacked by bombers and fighters since. USAF night bombing capabilities were limited. Only a handful of A-26K and B-57 bombers and one squadron of specially equipped F-4 fighters were night capable. Conventional fighter and bomber crews required illumination to operate at night, which in itself was a complicated tactical scenario. When C-130 *Blind Bat* and C-123 *Candlestick* aircraft dropped parachute flares to surveil suspect areas of the trail, PAVN drivers had time to pull off the roadway under jungle cover to avoid detection and attack. To better detect and attack traffic on the Ho Chi Minh Trail, new technologies were pressed into service. Operation *Igloo White* aircraft dropped Air Delivered Seismic Intrusion Detector (ADSID) sensors along the Trail near North Vietnamese Base Areas to detect seismic and audible signatures of enemy activity. Sensors impacting the surface would burrow into the soil, exposing an antenna that was built to appear to be a tree branch or plant. Raw sensor data was collected by EC-121R aircraft (call sign *Batcat*) while flying over a sensor string. *Batcat* then transmitted raw sensor data to the Infiltration Surveillance Center (ISC) in Nakhon Phanom (NKP) Thailand for processing. When activity was validated and mapped, the ISC sent targeting coordinates to Airborne Battle Command and Control (ABCCC) aircraft to direct strikes.

The task of choking off the Ho Chi Minh Trail was challenging from both tactical and political standpoints. An international diplomatic agreement prohibited US ground forces from operating in Laos. Though covert Central Intelligence Agency (CIA) operations and secret US military commando raids were conducted, the US government officially denied these operations occurred. Likewise, North Vietnam considered infiltration into Laos and Cambodia necessary, regardless of international law. Since neither side adhered to international law, Laos and Cambodia became a free-for-all for aerial attack of North Vietnam truck traffic by mining, bombing, strafing, and, of course, AC-130 strikes.

From November to June every year, monsoon conditions shifted, ushering in what was called the dry season. During this period, roads in Laos dried up, enabling large PAVN truck convoys to transport North Vietnamese troops and supplies into South Vietnam. This series of roads, trails, footpaths, and waterways originated near the Annamite Mountain Range of northern Laos. The first point of entry from North Vietnam was through the Mu Gia Pass, a strategic and well-protected choke point. Beyond the Mu Gia Pass, traffic branched out on a vast, meandering network running through the rugged mountainous terrain of Northern and Central Laos, the heavy jungle of Southern Laos and southward into Cambodia.

The most prolific trucks used by *Group 559* were Chinese ZiS-151 and Russian ZIL-157. These general-purpose, all-wheel drive, gasoline-engine vehicles were capable of hauling 2.5-ton payloads over rough terrain. The trucks were mechanically robust but unarmored and vulnerable to air attack. To mitigate detection and attacks from the air, truck drivers camouflaged their vehicles with freshly cut limbs and foliage. Though thousands of ZiS-151 and ZIL-157 trucks operated on the trail every night, normal convoys would range from five to fifty trucks, depending on the size of the Bihn Tram, time of year, and need for supplies. Every convoy was accompanied by a crane truck which would be used in the event a truck broke down or was damaged during an attack. Drivers knew their trails well, as did combat engineers and laborers who repaired roads and vehicles after attacks. Adjacent to most major roads, fox holes provided hasty driver protection from bomb fragments during an attack. Though trucks were a primary target for US aircraft, the PAVN made use of foot trails, bicycle paths, and even elephants to transport ammunition, food, and medical supplies southward.

The ZIL-157 truck was a primary target for AC-130 crews working the Ho Chi Minh Trail. (USAF Photo)

A large convoy transiting through the strategic choke point known as the Mu Gia Pass. (USAF Photo)

CHAPTER ONE

VIETNAM, LAOS, CAMBODIA COMBAT OPERATIONS, 1968–1969

Gunship II, aircraft 54-1626 in 1968. Note that the rectangular hole in the fuselage aft the 20mm guns was the original Infrared sensor location. (USAF photo)

Gunship II was deployed in February of 1968 to Ubon Royal Thai Air Force Base (RTAFB), Thailand, about forty miles west of the Laotian border to sustain the momentum of the AC-130 program until production aircraft could be delivered. The initial cadre of crews and maintainers included some of the same ASD personnel who participated in the combat evaluation deployment several months earlier.

Among the eclectic mix were pilots Major (Maj) Ron Terry who was a key developer of the AC-47D *Spooky* gunship, and Majors Jim Wolverton, Ken Gentry, and Jack Kalow. Navigators included Major Jim Krause and Captain (Capt) Chuck Meny and First Lieutenant (1st Lt) Vogler. Flight Engineers were Technical Sergeants (TSgt) Bob Davis and Gary Knecht. Loadmaster was Sergeant (Sgt) Fuller and Illuminator Operator was Master Sergeant (MSgt) Farris Heins. Gunners were Staff Sergeants (SSgt) Terrence Kenney and Paul Gillam. Mr. Bob Dunham was a civilian systems expert who flew on *Gunship II* combat missions to operate an experimental sensor. Wing Commander (Wg Cdr)[25] Thomas Pinkerton was a British exchange officer assigned to the Air Force Avionics Laboratory (AFAL) who designed and built the analog fire-control system. These men, along with other key personnel faced a seemingly never-ending set of challenges to establish a working capability. The first issue was command and control.

The sole gunship on station was assigned to Detachment 2 of the 14th Air Commando Wing, Nha Trang Air Base, Vietnam, but operated alongside four F-4 fighter squadrons of the 8th Tactical Fighter Wing at Ubon. Because of this command and control disconnect, deployment commander Lt Col Ross Hamlin faced difficult issues wrangling facilities and support. Most support arrangements were ad-hoc at best. Enlisted personnel and some officers were billeted in open-air barracks near the perimeter of base but some

25 The Royal Air Force rank Wing Commander is equivalent to Lt Col.

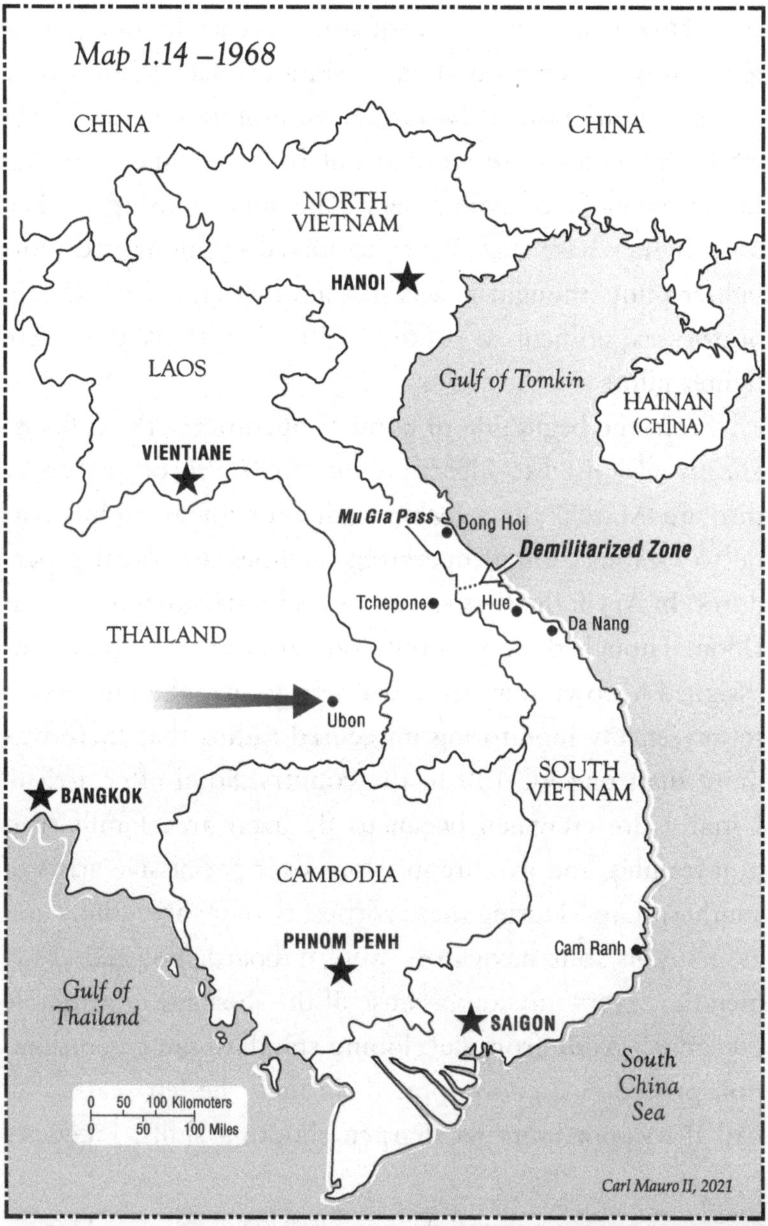

Ubon Royal Thai Air Force Base was located near the Laotian Border and central to some of the most active routes of the Ho Chi Minh Trail. (Drawing credit Carl Mauro II)

stayed in hotels downtown. Squadron-owned buildings were few and scattered around base. Maintenance facilities were virtually non-existent. Nevertheless, morale remained high while the issues were worked out by clever and resourceful crewmen and maintainers. The lone gunship parked on a fighter base was met with mixed opinions and most fighter pilots thought it was just another strange ASD laboratory experiment to make C-130 pilots think they were fighter pilots.

From the beginning of combat operations, the call sign *Spectre* was used to identify *Gunship II.* [26] From February through March, the gunship cadre fought an uphill battle to establish initial operating facilities and local procedures. In April, the first six pilots and navigators arrived at Ubon. Though there was only one aircraft, each pilot was assigned his own numbered call sign to give the impression to any enemy monitoring unsecured radios that there was more than one AC-130 in the country. Soon after arrival, initial cadre crewmen began to fly local area familiarization training and live fire missions over permissive areas of southern Laos. During these sorties, two or three pilots and as many as four navigators were onboard. Enlisted crewmen flew every mission because of the shortage of available personnel. Aside from developing standard crew coordination procedures, pilots rotated through the left seat so all had the opportunity to sharpen shooting skills. Likewise,

---

26 The origin of the squadron name *Spectre* has been lost in history.

sensor operators rotated positions, and if a third pilot was onboard, he augmented the NOD operator. This was a physically demanding duty since the NOD operator wore about thirty pounds of equipment and had to stand or kneel in an open doorway with 150 knots of wind blasting across the opening.

Call Sign	Initial Cadre Gunship II Pilots	Initial Cadre Gunship II Navigators
Spectre 01	Reserved for Commanders/VIP	Lt Col Alva Cross
Spectre 02	Lt Col Garvie Fink	Capt Tom George
Spectre 03	Lt Col Charles Koeninger	Capt Capt Hal Welch
Spectre 04	Maj Paul Zook	Capt Bill Tunnicliff
Spectre 05	Maj Maj. Bill Olsen	Capt Thurston Yoshina
Spectre 06	Capt Capt. Tom Sparr	Capt Cal Taylor
Spectre 07	Capt Capt. Bob Reneau	*Footnote[v]

*List is not all inclusive, but represents the only available information concerning the initial cadre. Information source: Spectre: The First Year, Col (Ret.) Cal Taylor.

Once crews were confident of their capabilities and training, they began to fly armed reconnaissance missions searching for traffic on the Ho Chi Minh Trail. Fire support

v. Crewmen and Maintainers included MSgt Joe Blanton, TSgt Bill McGee, SSgt Virgil Yarborough, SSgt Bob Dodson, TSgt Charles Beasley, TSgt Tom Bailey, Sgt Art Leath, Sgt Douglas Smith, MSgt Bill Spoor, SSgt Warren Coffee, SSgt Roger Eichler, Sgt Bill McGahey, Sgt Mike Kleiber, SSgt Ron Gagnon, SSgt Lance Timm, Sgt Jim Perry, SSgt Norm Grover.

for ground forces always took precedence over interdiction tasking, and the Direct Air Support Center (DASC), located in Bien Hoa, Vietnam, routinely requested *Gunship II* divert to support ground forces.

Sustaining *Gunship II* for an extended period was more than a challenge. The aircraft was a one-off prototype maintained by Jack-of-all-Trades maintainers to keep the basic aircraft in flyable condition. There were no USAF maintenance specialists assigned to the cadre and only a few *Blind Bat* C-130s on temporary duty at Ubon to borrow parts from. The many unique commercial avionics systems installed on the aircraft were maintained by civilian engineers from various manufacturers. Maintaining and repairing the unique systems of *Gunship II* was difficult at Ubon where no inspection facilities existed. The situation was complicated by a July through October monsoon condition commonly known as the rainy season. During this period, regular heavy rains made it impossible to complete phase inspections on the open flight line. An interim solution sent *Gunship II* to Naha, Japan every 150 flight hours for phase inspections, while a phase dock was being built on Ubon.

> "Frequent trips for phase inspections at Naha Air Base, Japan required us to remove everything from the aircraft and block the openings so we could pressurize. Since the old aircraft leaked like a sieve, we had a trash can full of water and a bunch of wet rags to block the pressure leaks. Once, at

> Naha, we needed a part that was unavailable so we ended up going to Futenma Marine Corps Air Station to cannibalize a part from a Marine C-130. We had a high priority and the Marines were not happy to see us, but we got the part. The same trip, we brought back a stray dog who became our mascot."
> **MSgt David Wilson,** Spectre Crew Chief

Missions over Vietnam were normally flown at 3,500 feet above ground level (AGL) while missions over the Ho Chi Minh Trail were flown at 5,500 feet. Because of these low operational altitudes, fuel consumption was high, so mission length would only be about 3.5 hours at the lower altitude and about four hours at the higher altitude.

Attacking trucks in blackout conditions took a great deal of skill, crew coordination, communications, and a pilot who had mastered traditional stick and rudder flying. Likewise, the copilot was skilled as the airspeed control officer who maintained constant airspeed since a few knots faster or slower than required would result in missed targets. Another important aspect of the attack orbit was on-nominal firing geometry. This aspect considered all environmental parameters including angle of bank, slant range to target, gun depression angle, ballistic trajectory, and such to project gunfire onto a target in the exact center of a nominal orbit. The pilot did all the firing and correcting. Effectiveness was highly dependent upon being in the correct geometry. Airspeed had to be exactly 150 knots. Angle of bank had to be 30 degrees which was continually called out by the flight engineer during the firing phase. In all cases, as soon as crews crossed into a hostile area, they would perform a live-fire boresight known as a "tweak" to ensure the gunsight and guns were harmonized. This procedure also confirmed guns were fully operational.

An on-nominal orbit could not be obtained until one of the two sensor operators located and tracked a target and its position was processed through the fire-control system which provided steering information and symbology onto a gun sight mounted on the pilot's left. Once the pilot superimposed his "pipper" onto the target, he directed the flight engineer (FE) to select any combination from one gun to all eight for an attack. Aerial gunners armed individual guns then serviced and reloaded them as needed once firing began. Normally, one pair of 20mm guns would be fired at a time, but the level of fire required depended on the tactical situation.

> "Missions were flown at 5,500' AGL (above ground level). The mini-guns were not effective at that range and were used only occasionally. Normal practice was to load one or more of them with tracer ammo to be used for marking targets. Beyond that, they were of little use unless conditions (absence of heavy ground fire) permitted operating at a lower altitude, such as working in support of ground troops." **Lt Col (Ret.) John Cahoon**, pilot.

While weapons effects were relatively straightforward and calculable, locating, tracking, and firing on a target was a challenge, especially at night. Night vision capabilities in the 1960s were an emerging technology that required comprehensive training and operators with deductive reasoning to be effective in combat. The IR detection set was state-of-the-art for its time and capable of detecting heat sources in complete darkness, but the image of a truck, bulldozer, or even an elephant was displayed as a relatively nondescript blob. The NOD was adapted from an Army ground system originally intended for short-range surveillance. It was a light amplifier that needed some level of ambient illumination to be effective and could not operate in complete darkness. If starlight, moonlight, or covert illumination of the 40KW searchlight was not available, the NOD was ineffective. AC-130 crews were hesitant to use the 40KW spotlight over Laos, so on dark nights, the IR was the only usable sensor.

Though the *Gunship II* crew set out to destroy every vehicle they encountered, the physics of doing so did not always agree, nor did the ability to prove or disprove damage claims. Standard BDA protocol established by the 7th Air Force included destroyed, damaged, and results not observed (RNO), which was sometimes used when a target was suspected of damage, but no other damage criteria could be applied.

20mm HEI impacts and explosions on a truck would produce a telltale flash of a hit, but there was no practical method to determine whether the truck was damaged to the point of attrition or if only cargo was affected and the truck was not damaged at all. These subjective parameters were questioned by both crews and the command structure, but both agreed a truck that caught fire was almost always destroyed.

Capturing BDA was challenging, especially at night. Early *Gunship II* missions had no ability to capture BDA except for an occasional press photographer, which was sketchy at best. Notwithstanding, the situation was complicated by the standard USAF BDA protocol of flying daylight photo reconnaissance missions the following day. This time lag allowed the PAVN time to tow damaged vehicles back to their Binh Trams for repair and push destroyed vehicles off the trail into the jungle for concealment. Seeking a better method to collect BDA, an aerial photographer was assigned to fly on *Gunship II* to capture real-time BDA on 16mm motion picture film by filming through the lens of the NOD.

Gunship II was assigned the responsibility to search and destroy vehicle and boat traffic on rivers, roads, and trails of *Steel Tiger East* and the *Tiger Hound* areas of Laos. A typical gunship mission started by searching for enemy movement west and northwest of Saigon. Crews searched the waterways and roads of III Corps and contacted various Army Special Forces camps as they arrived overhead. If needed, the camp's fire support officer (FSO) would request certain parts of their sectors be checked out.[27]

Missions flown on the Ho Chi Minh Trail were always under blackout conditions to avoid easy acquisition by

27 Unit Historical Records, 16 SOS, 1 Jul-30 Dec, 1968 (de-classified)

PAVN AAA gunners. Since the trail was protected by optically aimed 23mm and 37mm guns, which relied on tracers to aim, AC-130 crews quickly developed tactics to avoid being hit. Since it normally took about four to six seconds for a 37mm tracer to reach altitude, scanners had a few seconds to determine if AAA fire would be accurate or inaccurate, or if a defensive maneuver was needed. Tracers appearing to move fast across a scanner's field of view were deemed inaccurate, but tracers that appeared to be moving slowly (or stationary) were considered accurate since they were coming right at the aircraft. When an "AAA inaccurate" call was made over the intercom system, the pilot continued his attack orbit. When an "AAA accurate" call was made, it would be immediately followed by a break maneuver to avoid being hit. While the pilot watched for tracers on the left, a right scanner covered the right side and the illuminator operator (IO) covered the rear and underneath the aircraft. When accurate AAA was called, pilots quickly broke towards the direction of the gun, especially when the gun site was inside their orbit, to quickly negate an accurate trajectory.

The table navigator recorded the caliber and number of AAA rounds fired at them whenever crewmen announced AAA fire over the interphone system. The positions of gun sites were marked for intelligence reporting.

During the first four months of *Gunship II* operations, *Group 559* commanders took note of the number of trucks destroyed or damaged. Their response was to increase the number of 37mm AAA guns on the trail.

> "Mission effectiveness has been reduced significantly by 37mm ground fire. The 37mm fire is so intense that the gunships firing orbit must be discontinued after one burst. Where there are trucks there are 37mm positions. During the 57 sorties flown over the trails in June, the aircraft was fired at on 56 missions. The average number of rounds were 60, with 400 rounds being the high."
> 14th Air Commando Wing History, 1 Apr-30 Jun, 1968. (U)

On 14 June 1968, with concern over the increasing AAA threat in Laos and the onset of the rainy season, the 7th Air Force commander directed operations over the Ho Chi Minh Trail cease. *Gunship II* operations were then shifted to Tan Son Nhut, Vietnam to assist in combatting a rocket threat in the area. Upon arrival, Special Forces Major Don Faircloth began to fly as a member of the crew to help learn geographical aspects, landmarks and coordinate radio communications with ground units. Maj Faircloth fit in well with *Spectre* crews, referring to himself as "Casper the Friendly Ghost" to parody the gunship names *Spectre* and *Spooky.* One night, Capt Reneau's crew was cleared to fire on reported PAVN tanks in the Mui Bau Dhem Mountains. When Spectre 07 rolled in over the reported target coordinates, they located a few nondescript hot spots and questioned if they were indeed enemy tanks and felt their 20mm HEI would not do much damage even if they were hit. Assured they were tanks, Spectre 07 was cleared and opened fire, missing the targets by a large margin. While Spectre 07 continued the attack, high winds and steep terrain plagued the crew as they missed targets repeatedly. While Capt Reneau became frustrated with one missed burst after another, Major Faircloth stood in the left paratroop door, above the Bomarc radar system and began firing his service revolver at the tanks below.

> "We were all getting somewhat frustrated when Bob Dunham called on the intercom and said Bob, you're not going to believe this, but Casper has just emptied his service revolver out the left paratroop door trying to kill those tanks. Now that's frustration!"
> **Lt Col (Ret.) Bob Reneau**

After repeated attempts to hit the tanks with no effects on the target, Spectre 07 returned to Tan Son Nhut and called it a night.

Gunship II flew every night, searching for infiltrating enemy helicopters and landing only to refuel and rearm until the crack of dawn. *Gunship II* remained at Tan Son Nhut for several months, searching for Viet Cong 122mm rocket positions that were regularly fired onto US encampments. Finding the rockets was nearly impossible, however. The enemy was adept to field expediency and, instead of toting a launcher, they simply supported the rockets with two crossed sticks or laid them on the side of an embankment. Additional threats existed off base as well. *Gunship II* personnel were billeted in two hotels downtown which came under attack by the Viet Cong. These attacks normally occurred at night, when crewmen were flying a mission. *Gunship II* maintainers under attack rushed to the roof where weapons and ammunition were stored as South Vietnamese forces squelched the attack. Because of the threat level, all *Gunship II* personnel were moved onto base for security reasons.[28]

Missions were sometimes hampered by miscommunications between different controlling agencies. During a routine mission near the coast and just south of the Demilitarized Zone (DMZ) a large spotlight was directed at *Gunship II*. Unable to determine if the light was friendly, the crew elected to remain offshore and circumnavigate around the light to stay out of range of AAA range. The next morning, a USMC Major announced they had successfully thwarted penetration of the DMZ by a large aircraft the night prior. When it was revealed the large aircraft was a heavily armed Gunship, the Major got all huffy and said it was a good thing the aircraft didn't come any closer, or he would have shot it down with quad 50 caliber machine guns linked to the spotlight.

> "Our commander told the Marine Major it was a good thing they did not fire at us since we ate 50 cals for breakfast, which he, of course, did not appreciate. Fortunately, our Rules of Engagement prohibited us from firing at the spotlight. The sad thing was

28 Letter from Lt Col (Ret.) Bob Reneau

> because of fractured communications in general and the lack of complete knowledge of the whereabouts of friendly forces, accidentally firing on friendly forces was a very real concern from both air and ground perspectives." **Lt Col (Ret.) Bob Reneau, pilot**

One night, Spectre 07, commanded by Capt Bob Reneau, was flying a routine mission when directed to an encampment. The controlling agency was sure there were no friendlies in the area, but as *Gunship II* orbited waiting for clearance to fire, Sensor Operators noticed people in the camp were acting normally and did not take alarm at the presence of the gunship overhead. Additionally, there were no snuffed out fires, no vehicles moving, and no tracers coming from the camp. Spectre 07 questioned the status of the camp; nevertheless, the controlling agency directed the crew to fire. Capt Reneau reluctantly commanded the guns to be armed while refining his firing orbit. With only seconds to spare before firing, a frantic cease-fire call came over the radio. The controlling agency admitted they had made a mistake and the camp was actually friendly. This is just one example of a simple mistake that could have turned tragic with the push of a button.

Throughout the summer months, *Gunship II* continued to refine tactics and procedures in Vietnam at every opportunity. TIC missions were sometimes dicey, especially within the confines of villages. On one occasion, the mayor of a village was concerned about the possibility of Spectre 07 sinking his boat tied up on a dock behind his house. Unsure of the accuracy of the gunship, he asked the crew to fire on a large tree at a distance before clearing them closer to his boat. The mission was accomplished with complete success. Though new to Vietnam, *Gunship II* was building the same reputation as AC-47Ds did regarding accuracy.

While at Tan Son Nhut, *Gunship II* patrolled South Vietnam in support of US Army Special Forces camps, friendly hamlets, and outposts. South Vietnam was dotted with dozens of Special Forces camps, including large permanent camps near the borders of Cambodia and Laos. These camps were part of the Civilian Irregular Defense Group (CIDG) program. Each camp was commanded by a US Army Special Forces Captain or First Lieutenant and a small force of SF soldiers. Most camps had a dirt runway or at least a helicopter landing zone, a detachment of USAF Combat Controllers, and an Air Liaison Officer to support aerial deliveries. The bulk of each camp's population was comprised of local ethnic minority personnel, mostly Montagnards, who trusted neither North nor South Vietnamese governments.

Viet Cong forces routinely infiltrated in from Cambodia to target CIDG camps with mortars, RPGs, 122mm rockets, and small arms fire. One camp in particular—Katum, commanded by 1LT Donald Giddeon, 5th Special Forces Group, Detachment A-322—was attacked almost daily, earning the title "Kaboom." In most cases, attacks would last only a few hours before the Viet Cong were either defeated or withdrew. The Viet Cong favored periods of bad weather to conduct large-scale operations against CIDG camps since they knew aerial fire support would be limited.

On the night of 18 August 1968, poor weather encircled Katum camp. The 5th Viet Cong Sapper Battalion took advantage of the weather to mount an attack. Spectre 06, commanded by Capt Tom Sparr, was flying a routine hunter/killer mission when they were contacted by the DASC to support a TIC at Katum Special Forces Camp. When Spectre 06 arrived overhead, sensor operators located enemy personnel outside the perimeter and Capt Sparr began to engage them. Within an hour, Spectre 06 had started five fires and touched off four secondary explosions outside the perimeter. It was obvious the Viet Cong were armed with heavy weapons and explosives.

Eventually, the Viet Cong attack seemed to subside, but Spectre 06 remained overhead until about midnight when they had to return to base for refueling. Shortly after Spectre 06 departed, sapper squads advanced towards the outer perimeter wire. A call went out for help and an AC-47D *Spooky* was alerted. Once refueled, Capt Sparr switched seats with his copilot Capt Bob Reneau and the aircraft call sign was changed to Spectre 07. After take-off and a short flight to Katum, Spectre 07 arrived at a nearby holding point while *Spooky* continued to fire on advancing Viet Cong. Throughout the night, camp defenders fired small arms and point-blank 105mm canister rounds into the tree line to the north while Viet Cong mortars rained down on the camp. A little after 0300, Viet Cong Sappers broke through the wire and were met with a rain of bullets from *Spooky*. Now low on ammunition and fuel, *Spooky* departed while Spectre 07 established a firing orbit and continued to fire on the Sappers. Before the night ended, all enemy attempts to breach the wire were met with persistent *Spooky* and *Spectre* fire. Just about sunrise, the camp was secure, Viet Cong attackers withdrew, and *Gunship II* rolled out of orbit, headed for home. This was the first time *Gunship II* supported a major TIC mission in Vietnam, but it would not be the last.

> "While Searching Roads, III DASC called that Katum was under attack. Proceeding there, Katum asked that fire be directed at the wood lines west of the camp. Before the night was over, the gunship returned twice to Tan Son Nhut for refuel and rearm. At one point, Katum asked that fire be laid down on the west wall since the VC were trying to bridge the wire. This was the beginning of a great working relationship between the SF camps in III Corps and the AC-130." Unit History Report, 16 SOS 1 Jul-30 Dec, 1968. (U)

The support provided by *Gunship II* at Katum continued the AC-47D *Spooky* legacy, but with a higher level of lethality and sensor capability. The use of 20mm guns capable of firing high explosive incendiary (HEI) ammunition fortified the overall effectiveness of the gunship, but the legacy of fixed wing gunship fire support for friendly forces that was started by the AC-47D *Spooky* remained the same. *Spectre* had established its place alongside *Spooky* and, nearly every night, both models patrolled multiple outposts, camps, and hamlets looking for guerilla activity while supporting friendly forces.

On 22 August 1968, *Gunship II* was tasked with searching the Demilitarized Zone (DMZ) for enemy helicopters infiltrating the enemy into South Vietnam. Though there were no credible sightings of enemy helicopters, Intelligence sources suspected North Vietnamese helicopters were flying at night to avoid being spotted during the daytime. At the time, there was virtually no capability for fighters to detect and engage blacked-out aerial targets at night. Crewmen considered the task unrealistic since was difficult enough to attack a slow-moving ground target, let alone a fast-moving aerial target flying at an unknown altitude. Nevertheless, *Gunship II* began flying all-nighters averaging twelve hours per mission, landing only to refuel. During one refueling at Da Nang, the field was mortared prompting the closure of the field. After consulting with base headquarters, the gunship crew was authorized to launch to search for aggressors. *Gunship II* remained on station until about 0830, then returned to Tan Son Nhut. Two nights later, another mission to search for helicopters along the DMZ was diverted to provide fire support for three separate ground units. Again, the crew flew an all-nighter, logging twelve hours flight time.

Just over a month after the overrun attempt at Katum, *Gunship II* provided its second critical fire support mission in support of friendly forces. On 25 September 1968, Capt Tom Sparr (Spectre 06) took off on a routine armed reconnaissance mission searching for enemy activity in Vietnam near the Cambodian border. Co-pilot Capt Bob Reneau noted weather was good at 3,500 ft. AGL as they patrolled roads and trails. Lt Col Koeninger (Spectre 03), the squadron operations officer, was the third pilot. Just about the time Spectre 06 was low on fuel, they received a distress call from the US Army Captain William E. Clark, commander of US Army Special Forces Detachment A-323 at camp Thien Ngon. This major fighting camp was located in a free-fire zone about five miles from the Cambodian border and was a location routinely patrolled by *Gunship II* crews. Like Katum, Thien Ngon was star-shaped and easy to identify from the air. The outer perimeter consisted of *fougasse*, trenches, anti-personnel and anti-tank mines, and barbed wire. Located on each star point were sandbag-protected fighting positions, just inside spools of extended barbed wire. The inner compound contained an arrangement of large metal shipping containers (Conex), protected by sandbags, surrounded by a pentagon-shaped moat filled with stakes and barbed wire. There were three Vietnamese and two Cambodian CIDG companies assigned to the camp, which was supported by US Army Special Forces soldiers under command of Captain Clark and executive officer 1LT Mike Winslow. Supporting the airlift element of the camp were several USAF Combat Controllers including TSgt Mort Freeman and Air Liaison Officer Lt Col Charles Wurster, a C-130 pilot.[29] There was at least one patrol outside the wire

29 Lt Col Charles Wurster was the father of Lt Gen Donny Wurster, who eventually became Commander, Air Force Special Operations Command from Nov 2007-June, 2011.

at all times searching for PAVN activity, that almost always originated from across the border in Cambodia.

> "Our camp was under attack for 30 days. It began around midnight on September 1, when over 500 rounds of 122mm rockets and 82mm mortars tore our camp apart. Every major structure was hit. A 122mm rocket and an 82mm mortar came through the team house roof wounding four of us and killing the team medic. We were very fortunate that the enemy did not launch a ground attach that night. For the next 29 days, the enemy fired rockets and mortars and dug trenches towards the camp, along with limited ground attacks. The enemy had 12.7 caliber anti-aircraft weapons to defend themselves from air attack." **1LT Mike Winslow,** executive officer, US Army Special Forces, Thien Ngon, 1968.

The camp had been under siege for nearly a month; but, on 25 September, a large enemy force was detected assembling nearby. That night, as Spectre 06 neared Thien Ngon, Capt Clark advised Capt Sparr the enemy was fast approaching and they expected a big firefight. When Spectre 06 arrived overhead, they were low on fuel and did not have much time to search. Pulling off target, the crew flew to Tan Son Nhut for fuel and ammunition. Just before Spectre 06 got out of radio range, Capt Clark radioed a message that enemy forces were probing the perimeter and they needed help quickly. With little fuel remaining and a failed generator, Capt Sparr landed at Saigon. In a short time, Capt Reneau and the flight engineer replaced the generator while the rest of the crew refueled and re-armed the gunship. Once ready, Capt Reneau (Spectre 07) took command and the crew took off towards Thien Ngon. The weather had changed since their earlier sortie as Spectre 07 navigated around thunderstorms en route to Thien Ngon. When Spectre 07 arrived overhead at 3,500 feet, it was obvious a full-scale attack had begun. Muzzle flashes dotted the perimeter, mortars were exploding inside the camp and trip flares lit advancing Viet Cong who were maneuvering towards the perimeter.

> "When we arrived at the base, the fire fight had fully developed. Tracer rounds were coming from the compound, trip flares were going off around the outer perimeter defense, mortar rounds were going off inside the compound. We were able to pick up some muzzle flashes from small arms and or mortars, but for the most part, we had to be selective in targets, not knowing where the friendlies were located. I asked if they had any "ball games" out (a term for patrols). They had, but were fairly confident in describing their locations." **Lt Col (Ret.) Bob Reneau**

Because of the close proximity of friendly and enemy forces, Capt Reneau was aware there was a high potential for fratricide when engaging advancing Viet Cong on the perimeter. Almost immediately after establishing a firing orbit overhead, enemy activity subsided as if they were waiting for the gunship to depart. Feigning departure, Capt Reneau rolled out of firing orbit to give the enemy the impression Spectre 07 was departing. About five minutes later, Capt Reneau made a 180-degree turn and headed back to Thien Ngon. Rolling in overhead, sensor operators Lt Col Cross and Capt Tunnicliff detected teams of Viet Cong sapper squads charging the barbed wire with Bangalore torpedoes in hand. While the sappers attempted to push the explosive tubes into the perimeter wire, Capt Reneau confirmed all friendly personnel were undercover, then opened fire with all four 7.62mm guns. Thousands of bullets rained onto the sappers, killing them on the spot.

> "I pulled off, as if leaving the area. I waited about five minutes and returned to what appeared to be a quiet battlefield. Fortunately, the ploy worked...you can guess what happened next...talk about a rain of fire. I completely saturated the area." **Lt Col (Ret.) Bob Reneau,** pilot

For several hours, wave after wave of sappers attempted to breach the wire but suffered the same fate as the previous wave via fire from Spectre 07's guns. When *Gunship II* was low on fuel, an AC-47D arrived on station to replace them. Since *Spooky* was blacked out at the same altitude, Lt Col Koeninger turned on *Gunship II*'s anti-collision light, which was evidently seen by a Viet Cong gunner who engaged and hit *Gunship II* with 12.7mm machine gun fire. Undeterred, Capt Reneau followed tracers to the site then targeted and killed the gunner before departing for low fuel.

> "I remember the red tracer rounds arching up towards Spectre 07 and thought the gunner was either crazy or one hell of a hero, as Spectre 07 unleashed a torrent of rounds upon him." **1LT Mike Winslow**, camp XO.

After handing off duties to *Spooky*, Spectre 07 returned to Saigon for rearming and refueling. Since the fifty-caliber hit was nothing more than sheet metal damage, the aircraft was ready for another mission to Thien Ngon. Once again, pilots rotated positions and Lt Col Koeninger (Spectre 03) took off for their third sortie. When they arrived over the camp, attackers had withdrawn from the perimeter and were on the run. After a brief search, Spectre 03 sensor operators detected them crossing the border into Cambodia. Since Rules Of Engagement (ROE) prohibited US forces from entering Cambodian airspace, ground control radar tracked Spectre 03 as Lt Col Koeninger neared the border. Nevertheless, Lt Col Koeninger broke the rules, crossed the border, and engaged enemy personnel, killing a number of them.

Though three sorties were flown by three pilots, Capt Reneau undoubtedly broke off the attack at Thien Ngon on 25 September, thus saving friendly forces from being killed or captured. All told, the gunship fired 6,000 rounds of 20mm and 12,300 rounds of 7.62mm that night. Many lessons were learned, especially the importance of tactical deception and the element of surprise. Post-mission fallout of Spectre 03's mission resulted in an international incident whereas the government of North Vietnam claimed Spectre 03 killed innocent Cambodian villagers. An investigation, however, revealed the personnel were actually Viet Cong guerrillas dressed to appear as Cambodian villagers. Fortunately, the crew was vindicated, but illegal border crossings continued to be frowned upon. Enemy efforts to seize Thien Ngon ended five days later after the decimated force made one final push into the perimeter wire littered by decomposing bodies of their dead comrades, but again, they failed to overrun the camp.

> "The final ground assault was on the night of September 30. The enemy fired over 1,000 rocket, mortar and B-41 rounds into the camp. The enemy sappers cut lanes through the barbed wire and soon 500 to 700 enemy attacked the camp throughout the night. The next morning, I counted 150 enemy bodies around the perimeter of the camp. If it were not for the brave and resilient aircraft crews of the fast-movers, Spooky and Spectre gunships, the result might have turned out different." **1LT Mike Winslow,** camp XO.

After-action analysis revealed the force that attacked Thien Ngon was two 9th Viet Cong Regiments estimated to be 1,000–2,000 men. Previous to the night of 25 September, they attacked relentlessly with mortars and rockets almost daily for nearly a month to soften camp defenses. Their end goal was to breach the wire, seize the camp, and kill or capture everyone within. Fortunately, the crew of Spectre 07 was in the right place at the right time with the right capabilities to kill multiple waves of sappers before they were able to breach camp defenses.

While combat activity continued in Vietnam into the fall of 1968, the gunship modification program was behind schedule and only four of the seven projected AC-130s were delivered to the USAF. Two were stationed at Lockbourne, Ohio in the 4413th Combat Crew Training Squadron (CCTS) and two were being readied to be sent to SEA. Coinciding with late aircraft deliveries, only three crews had been trained. This was not what the 7th Air Force commander wanted to hear since an especially active dry season in Laos was predicted. As a result, all four production AC-130s were directed to be deployed to Ubon for combat duty. Basic gunship training at Lockbourne then shifted to the AC-119 *Shadow* gunship until more AC-130s were delivered.

Back at Ubon, October weather was favorable for *Gunship II* TIC missions. Between 3–8 October, four special missions were flown in support of the besieged camp at Thuong Duc, South Vietnam. Suspected and known enemy positions were hit as directed by ground commanders resulting in six buildings destroyed, seven secondary explosions observed and four fires started.

By the fall of 1968, US President Lyndon B. Johnson faced political pressure due to the Tet Offensive and failed US strategy of bombing North Vietnam. The new strategy was to sever the Ho Chi Minh Trail to effectively choke off all supplies, equipment, and personnel flowing from North Vietnam through Laos and Cambodia into South Vietnam. On 11 November 1968, President Johnson halted all bombing operations north of the 20th parallel. This order shifted hundreds of aircraft from North Vietnam to the Ho Chi Minh Trail. Truck parks, weapons caches, and support facilities located under jungle canopy were difficult to detect, but vehicles spotted on open roads and trails during daylight were quickly targeted and attacked. During the daytime, low and slow FAC aircraft with long loiter time capability searched for targets like a hunter waiting for prey to appear. When they located targets, they called in fast-moving jet fighters, to bomb them. Overall results were generally satisfactory during the daytime, but during the hours of darkness, the capabilities of the FAC were limited to handheld starlight scopes, similar to the AC-130's NOD. For this reason, the PAVN discontinued daylight convoys in favor of nighttime movements using the cover of darkness to avoid detection. This shift in strategy pitted the USAF and USN with the difficult task of finding thousands of camouflaged trucks, hundreds of storage areas, and over one thousand miles of travel routes hidden in dense foliage and triple canopy jungle in Laos.

After the *Group 559* commander directed truck shipments only occur during the hours of darkness, the small fleet of night capable fighters and bombers bombed known strategic choke points and intersections hoping to either slow or stop truck traffic. The trail network was vast and complicated, yet despite bombing efforts, PAVN combat engineers and laborers rapidly repaired bombed trails or built detours around damaged areas. Earlier in the year, 7th Air Force commander Gen William W. Momyer was greatly impressed with *Gunship II* capabilities. Coincidentally, it didn't take long for the HQ PACAF staff to realize the AC-130 was the most capable aircraft available to hunt trucks on the Ho Chi Minh Trail. Not only could it find and destroy trucks at night, but it could also remain airborne for many hours, searching for targets of opportunity and did not need a FAC. Because of these factors, and the proximity to the Ho Chi Minh Trail in *Steel Tiger East*, Ubon Royal Thai Air Force Base was chosen for the AC-130's permanent home base.

On 31 October 1968, the 16th SOS was activated under the command of Lt Col Garvie Fink. The squadron was assigned to the 7th Air Force, Pacific Air Forces, 8th Tactical Fighter Wing, Ubon, Thailand. During the ceremony, the unit's sole gunship was on temporary duty status at Tan Son Nhut to support the TIC mission in Vietnam. By early November, two production AC-130s (54-1629 and 54-1630) arrived at Ubon. By that time, the squadron had grown to forty-four officers and ninety-six airmen, and more aircraft and personnel were expected before the end of the year.

On 24 November 1968, the first large-scale effort to interdict the Ho Chi Minh Trail, Operation Commando

Hunt, began.[30] To better control the large amount of air traffic expected over Laos during the operation, the eastern half of the country was sectioned into Visual Reconnaissance (VR) areas. Considering the priority of interdiction of the Ho Chi Minh Trail, the 7th Air Force commander verbally directed *Gunship II* to return to Ubon to participate in the operation. Ubon was closely located to some of the most active routes of the Ho Chi Minh Trail, specifically *Steel Tiger East* VR-4A and VR-4B near Tchepone, (also known Xépôn and Sepon) Laos, and Base Area 604, the headquarters of *Group 559*.

When 16 SOS leadership and crews were briefed on their primary interdiction mission, confidence was high, but crews and maintainers knew there were potential troubles afoot. The squadron had grown fast; perhaps too fast. Most of the crews arriving from Lockbourne were hastily trained at a minimal skill level because the demand to field qualified crews outweighed available training time. Combat experienced instructor pilots at Lockbourne included *Gunship II* instructor pilots Majors Jack Kalow and Ken Gentry and navigator instructor Major Jim Krause. Unfortunately, the demand for training exceeded what could be provided by the *Gunship II* cadre. To meet the demand, experienced C-130 pilots and navigators without gunship combat experience were pressed into service.

> "A goodly amount of our training time was spent in undoing some bad habits they picked

30 Commando Hunt ran from 1 Nov, 1968-30 Apr, 1969.

> up at Lockbourne from non-combat experienced pilots. In fairness, Jack Kalow and Ken Gentry were instructing at Lockbourne, however the output was so large that some non-combat instructors were used and we could really tell the difference in combat as to which ones had which instructors." **Lt Col (Ret.) Bob Reneau,** pilot.

Though experienced USAF navigators (known as table navs) were selected for the AC-130 program, they were not required to be qualified in the C-130 but were required to be highly experienced instructors or evaluators. This was especially important since navigator duties on the AC-130 were extensive and challenging. Not only were navigators required to perform navigation from point to point, but to guide their crews to specifically defined areas, roads, and trails while using equipment not specifically designed for the task. The table navigator's tools were maps, charts, radar, the E6B flight computer (aka "whiz wheel"), and a large dose of intuition. Aside from the table navigator, two navigators operated NOD and IR sensors, which were not used anywhere else in the USAF. Without doubt, navigator training on gunship-unique systems was difficult. Flight engineers were required to be highly experienced instructors or evaluators selected solely from the C-130 program. Illuminator operators were experienced C-130 crew chiefs on flying status and gunners were volunteers already trained as weapons loaders or gun shop technicians who were experienced with 20mm and 7.62mm guns.

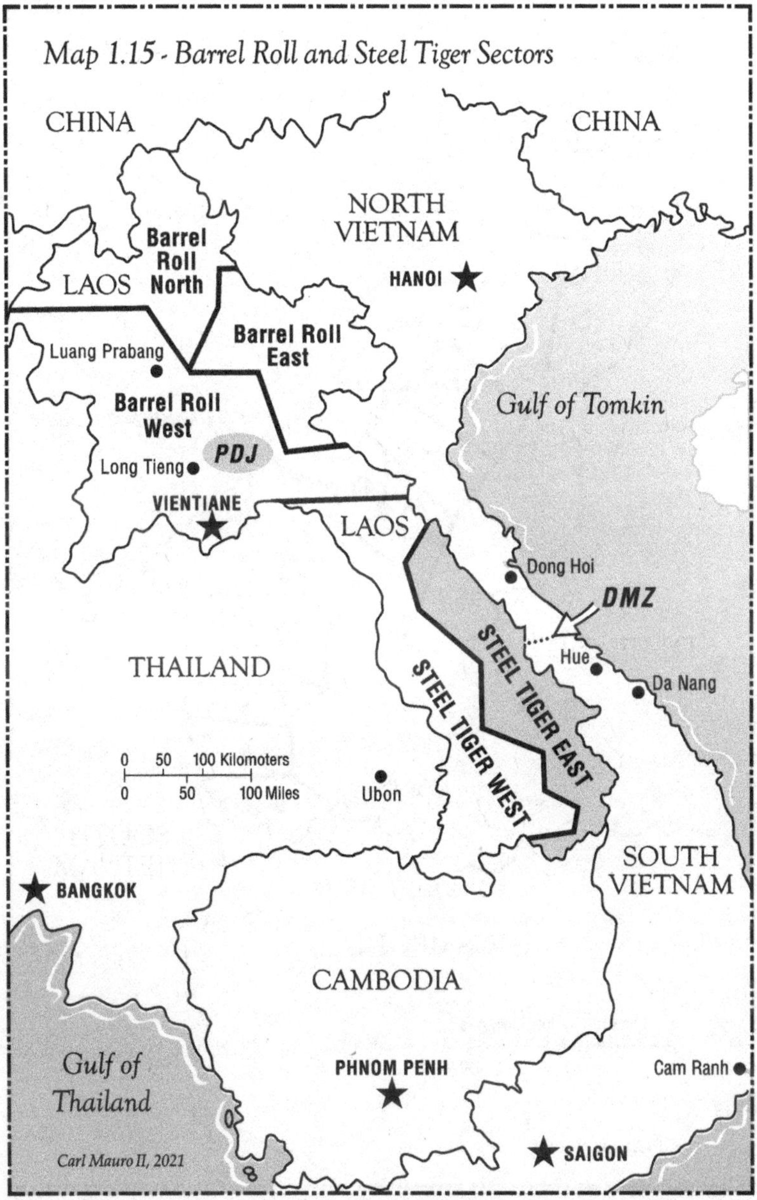

Map 1.15 - Barrel Roll and Steel Tiger Sectors

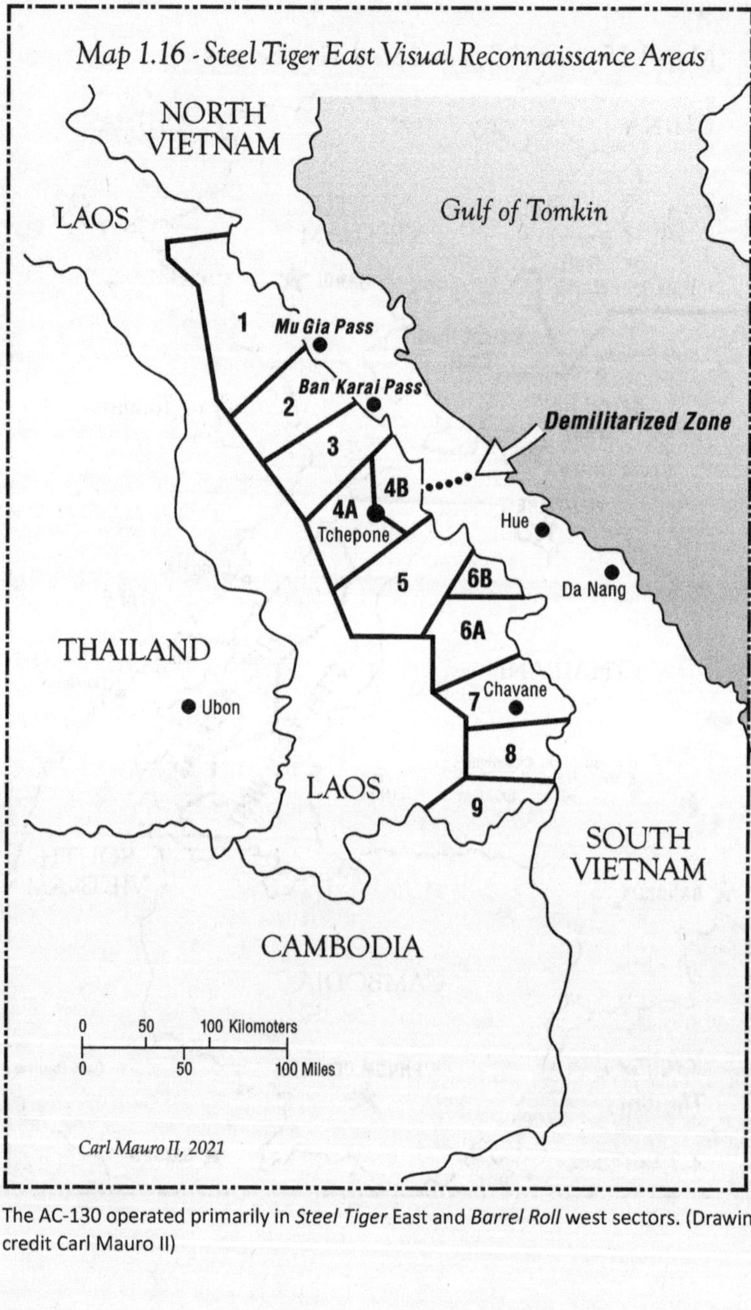

The AC-130 operated primarily in *Steel Tiger* East and *Barrel Roll* west sectors. (Drawing credit Carl Mauro II)

By late November 1968, the first three production AC-130s were stationed on Ubon. One more was scheduled to arrive in early December, and three more were being modified at Ling-TEMCO-Vought (LTV) in Greenville, Texas. Eager to begin *Commando Hunt* operations, newly arrived crews were briefed on local procedures and given a few local missions to familiarize themselves with the area. A standard landmark used for sensor alignment training was a well-lit fountain in the middle of a traffic circle in downtown Ubon. Next came flights into the relatively permissive "golf" VR sector of *Steel Tiger West* in Laos known as "the Monkey farm." There, AC-130 crews rarely encountered AAA or traffic before transiting into the heavily-contested *Steel Tiger East* VR sectors.

With production AC-130s arriving on station, it was realized that the *Gunship II* prototype was configured vastly different than production models and needed a major avionics upgrade. With the first three *Plain Jane* aircraft on station at Ubon, *Gunship II* was ordered to return to Wright-Patterson AFB.

> "Gunship II flew its last combat mission on 12 November, 1968. From February-November, Gunship II flew 148 combat missions and was credited with destroying 228 trucks, 32 boats, five buildings, three 37mm guns, two ZPU guns, five .50 caliber guns and two mortars. It also damaged 133 trucks, 8 boats, started 140 secondary

> fires, caused 97 secondary explosions, and was credited for 225 enemy killed. The aircraft had fired 565,900 rounds of 20mm and 463,400 rounds of 7.62mm and 1,601 MK-24 flares. An impressive record not only for the aircraft, but for the crews who flew dangerous missions in spite of heavy AAA and multiple avionics malfunctions." 16 SOS Historical Records, Dec, 1968.

By the second week of December 1968, all four AC-130A *Plain Jane* gunships were on station at Ubon and fully involved in interdiction efforts on the Ho Chi Minh Trail. This demanding mission required the 16 SOS to fly three or four missions per night, seven days a week. Crewmen usually flew two nights on, one night off, but it was common for some to fly more often to fill vacancies related to shortages or crewmen grounded for medical reasons. Basic mission profiles, however, were essentially identical.

Once production aircraft began operations at Ubon, a laundry list of problems created by LTV was identified. These problems were known and reluctantly accepted by ASD during aircraft modification to keep the program on schedule. Problems included spent casing bins that were too small, control boxes located in awkward places, black-out curtains were required to keep light from the cockpit from affecting the NOD, ammunition cans needed to be relocated, and the aft mini-gun shelf was too short. A list of discrepancies was compiled by Lt Col Fink, who sent a deficiency report to ASD requesting relocation or modification of a number of subsystems. Nevertheless, these items were considered growing pains, since the aircraft performed well and shot tighter groups than the prototype aircraft. As long as solutions to the aforementioned problems were being worked, 16 SOS leadership saw no need to stall the program.

During initial operations at Ubon, many other problems were encountered. Avionic malfunctions were sometimes crippling, the fire-control computer and IR set were relatively unreliable, and several other systems had high failure rates. Most of the time the fire control computer didn't work at all, so pilots simply used "Kentucky windage" while firing short bursts of 20mm HEI to walk fire onto targets. Crews often took off with an inoperable fire control computer since they were confident a work-around solution would be figured out on the way to their working area. There were few missions where all systems worked as they should, so crews learned to adapt and overcome minor issues. Major breakdowns, however, caused frustration, missed targets, and exposure to AAA fire with no ability to effectively return fire. If all attempts to remedy a problem failed, the crew's only available option was to return to Ubon for repairs.

Maintainers were equally challenged. The speed of development of the AC-130A and the state of some of the oldest C-130s in the inventory resulted in configuration differences between aircraft, especially in wiring and avionics. Crew Chiefs had to accept the fact that each gunship had its own specific personality and adjust accordingly. To top it off, technical data on nearly all the gunship modifications was essentially nonexistent. Maintainers dealt with many head-scratchers while troubleshooting complex problems.

> "Nothing was easy and changes were constant, but those were my birds and I loved the challenge. We made our own special tools, e.g. nipped and bent box end wrenches for removing components inside the cowling. We adapted and overcame a shortage of support equipment by using our web belts to wrap around the alternators and generators to physically suspend them while removing or installing. We worked around the clock to keep the AC-130s flying." **Sgt Carliss Abejon,** Spectre Electrician.

Maintenance and operations facilities at Ubon were ad-hoc at best. Systems engineers and commercial industry technicians made numerous trips to Ubon to troubleshoot complex avionics issues. Since the AC-130's unique systems relied heavily on contractor support, command leadership realized USAF maintenance technicians could not take over without training and technical manuals. Though some headway in training and documentation efforts continued throughout the latter half of 1969, it did not eliminate the need for contractor support. The demands put on maintainers to keep up with equipment changes was extreme and sometimes as soon as they were proficient with one system, it was replaced by another. Regardless of training and maintainability constraints, *Commando Hunt* tasking continued to challenge *Spectre* crews and maintainers at a high level of effort.

Though many other aircraft were involved in *Commando Hunt* during daylight hours, the AC-130 and one squadron of F-4s were the primary night-time truck hunting aircraft. Daytime fighter and bomber tactics varied, but they most often attacked strategic road intersections with delayed fuzed bombs, to make intersections impassable. Trail workers were quick to build camouflaged out-of-the-way detours far away from bombed-out intersections to keep trucks rolling. Though *Group 559* drivers may have believed they were less likely to be attacked by the AC-130 while on a detour, the opposite was true since detours routed trucks through less defended areas. Once channeled into a detour, AC-130 crews were able to locate convoys with their IR sensor and fire on the convoy without receiving any AAA at all.

The PAVN 591st Anti-Aircraft Artillery Regiment had battled against the AC-130 since the earliest days of *Spectre*. Equipped with 37mm AAA guns, they fired thousands of rounds at AC-130s but failed to shoot one down. The number of 37mm AAA traps was continually increased to heighten potential of downing an AC-130. These traps were composed of multiple guns spaced about a kilometer apart, but all aimed into the same kill zone in the sky. PAVN AAA trap tactics mandated AAA gunners hold their fire until an AC-130 entered a pre-planned multi-gun kill box in the sky. When an AC-130 would enter the kill box, all guns would fire simultaneously and aggressively at the gunship until the aircraft was either hit or exited the kill box. *Spectre* crews called the tactic "Dr. Pepper fire" to parody the soft drink slogan, "10-2-4," meaning AAA fire came from left, right, and behind at the same time.

In response to heavy AAA on the trail, *Spectre* launched a surprise of its own. No longer would a sole gunship fly orbits overhead while AAA guns continuously fired at them with near impunity. Instead, night capable F-4 fighter escorts from the 497th Tactical Fighter Squadron *Nite Owls* accompanied each AC-130, ready to drop bombs to suppress or kill gun sites. Forward Air Control (FAC) of the fighters was the responsibility of the AC-130 Aircraft Commander and both AC-130 and F-4 escorts operated in complete blackout conditions. To prevent a mid-air collision, a single shrouded light called "the canned beacon" only visible from above was installed on every AC-130. Keeping fighters paired up with AC-130s was in itself a task since the gunship could loiter for long periods over a target, but the F-4s couldn't. To sustain effective escort coverage, three F-4s were normally assigned to one gunship. During routine operations, one F-4 was "on the perch" at all times while the other two were either coming from or going to an aerial refueling track over Thailand. The AC-130/F-4 lash-up worked well since the fighters flew at about twice the altitude of the gunship and PAVN gunners did not make the connection it was a hunter/killer team overhead until it was too late.

Throughout the dry season, threat conditions on the Ho Chi Minh Trail continually escalated. Whenever AC-130s hunted a new section of the trail, the PAVN would move AAA guns into the area. *Spectre* crews and North Vietnamese AAA crews continually altered tactics to gain an advantage over the other. Because of the rapid build-up of AAA sites on the trail, many gun sites were manned by inadequately trained gunners who fired wildly at the AC-130. *Spectre* crews found these inaccurate sites easy prey and began to fire on them. Unfortunately, these duels were distractions to the gunship's primary interdiction mission, and time spent engaging AAA sites allowed trucks to escape unscathed. As a result, the 16 SOS commander directed Rules Of Engagement (ROE) prohibiting AC-130s from dueling with AAA sites. From that point on, primary AAA suppression was the responsibility of F-4 escorts and AC-130 crews would rarely engage an AAA site and if it did, only if it were easy picking on the inside of their firing orbit.

> "In most instances, particularly in an area of concentrated AAA it was advantageous to use F-4 aircraft for escort to suppress any offending gun site. Utilizing the team of the AC-130 and F-4 escort the gunship was normally able to remain on target for longer periods of time. When ground batteries would open fire on the gunships, the fighter escorts orbiting above, behind, and generally to the right of the AC-130 would dive to attack the position. Usually the F-4 pilots would see AAA fire before the gunship crews did and if so they would call and the gunship would clear them in. Sometimes of course,

> evasive action was required on the part of the AC-130 to escape AAA fire and the F-4 had to wait until the AC-130 was clear of his attack path before diving. " 16 SOS Historical record, 1 Jan-31 March, 1969. (U)

As *Commando Hunt* entered its second month, PAVN gunners adapted their tactics to counter escorted AC-130s. To mitigate exposure and make attacks more difficult, gun sites were located on high terrain, camouflaged with local vegetation and each gun crew was assigned a particular box of sky into which to fire into. This helped conceal the gun and prevented the need to clear the jungle for 360 degrees around the site. Gun crews would only fire when an AC-130 passed into a gun crew's assigned box and would stop firing as soon as the gunship departed their box. This tactic not only kept the gun concealed but simplified engagements and firing solutions. When viewed from the AC-130/F-4 perspective, this made it more difficult to pinpoint a AAA site, since the only time their location was visible was when tracers were spotted emerging from the jungle. To better enable successful suppression, the AC-130/F-4 team developed a simple and effective solution to pinpoint AAA locations. The tactic used MK-6 ground marker flares, which were essentially wooden logs filled with slow-burning compounds similar to what was used in common road flares. When a AAA site began firing, the IO would drop one MK-6, then another one 180 degrees from the first as to bracket the gun position. Fighters then used the two points of light to establish a run-in heading to attack the AAA site. The same tactic was often used to mark each end of a truck convoy to give F-4 pilots a visual reference for delivering their ordnance.

By January 1969, AC-130 interdiction missions on the Ho Chi Minh trail were tasked at an all-time high. Though it was considered the dry season, the weather was not perfect by any stretch of the imagination. Low cloud decks hampered crews on many missions, but when a crew was diverted for weather on 4 February, they were surprised to find a major PAVN staging area.

That night, Lt Col Thomas Simone flew *The Arbitrator* (call sign Spectre 01) to a river near Laotian Route 8 and Delta Point 75. Both FLIR and NOD operators detected 21 large boats about 40 feet in length that were being used to transport supplies across the river. Several were tied together in what appeared to be used as a floating bridge. The FLIR operator, Capt Larry Flinn, detected four small trucks on a trail near the shoreline so Spectre 01 rolled in and systematically destroyed all four trucks, then fired on the group of boats, but even when hit, they did not explode or sink. During this lengthy attack, a single 37mm AAA site fired at Spectre 01. In response, the gun site was bombed by an F-4 escort.

A few days later, *Mors De Caelis* commanded by Capt Maurice "Mauri" Galey, defended a Laotian outpost at Ben Thateng against North Vietnamese troops. *Spectre* crews also destroyed 48 trucks in two missions on 22 and 23 February; a truck kill rate higher than any other weapon system working the Ho Chi Minh Trail.

In late February, 7th Air Force Commander Gen George Brown visited Ubon for a closer look at the aircraft and crews who were turning in such spectacular results. During his visit, he viewed BDA film and intelligence reports of AC-130 missions. When he returned to his HQ, the 16 SOS received a congratulatory message:

ZNY EEEEE ZOK JPCCO
P 260210Z FEB 69
FM 7AF TSN AB RVN
TO RUBCTK/8TFW/C/UBON RTAFB THAILAND
INFO 7/13AF UDORN RTAFB THAILAND
BT
UNCLAS E F T O
FOR COL PATTILLO FROM GEN JONES
SUBJECT: OUTSTANDING ACHIEVEMENT (U)

1. I REQUEST THAT YOU EXTEND MY PERSONAL CONGRATULATIONS TO THE SPECTRE AIRCREWS OF THE 8TFW FOR THEIR OUTSTANDING RESULTS IN THE NIGHT TRUCK KILL PROGRAM.
2. THE BDA ATTAINED BY THE CREWS OF THE SPECTRE AIRCRAFT HAS BEEN AND CONTINUES TO BE EXCEPTIONAL THIS COULD HAVE BEEN ACCOMPLISHED ONLY THROUGH OUTSTANDING AIRMANSHIP AND SUPERIOR COORDINATION OF THE AIRCREWS AND THE EQUALLY

OUTSTANDING GROUND CREWS. ALL MEN IN THE SPECTRE SQUADRON CAN TAKE GREAT PRIDE IN THE FACT THAT THEIR EFFORTS HAVE PLAYED A LARGE PART IN OUR INTERDICTION PROGRAM AGAINST THE ENEMY.

3. AGAIN, I EXTEND MY SINCERE AND HEARTIEST CONGRATULATIONS TO THE AIRCREWS AND MEMBERS OF THE SPECTRE SQUADRON FOR A JOB WELL DONE.

PAGE 2 RUSVS 44IS UNCLAS EFTO

I HOPE THE EXCELLENT RESULTS OBTAINED IN PAST PERFORMANCE WILL BE CONTINUED IN THE FUTURE. I RECOMMEND THAT THE APPROPRIATE RECONGNITION BE GIVEN TO THE SPECTRE AIRCREWS FOR THEIR OUTSTANDING ACCOMPLISHMENTS.

BT (U)

The damage *Spectre* crews were inflicting on *Group 559* exceeded anything seen before. *Group 559* truck drivers who survived attacks came back to their Bihn Tram with stories of fear and anguish. In response, North Vietnam bolstered their 591st Anti-Aircraft Artillery Regiment defenses by adding even more 37mm AAA gun sites in *Steel Tiger East*. *Spectre* crews adapted quickly by operating between known AAA gun sites, just out of range of PAVN guns. These sites were marked on navigator's charts enabling crews to orbit for long periods of time, patiently waiting to ambush trucks as they passed on the trail beneath them.

> "Operations conducted in an AAA environment required two essential conditions be known, the exact position of the aircraft and the precise location of all AAA sites. With these factors known, the AC-130 was able to patrol an area where the AAA emplacements were as close together as six or seven miles. One procedure used in areas of known heavy AAA was to wait until the truck or target traffic moved away to a more permissive location before launching a strike." 16 SOS Historical record, 1 Jan-31 March, 1969. (U)

Regardless of *Spectre's* tactics to avoid AAA fire, AC-130s did get hit. On 3 March 1969, Maj Glen Franklin's crew was sent to the Delta 75 region of Route 8 in central Laos. After entering a firing orbit, a 37mm projectile hit their right elevator. The pilot maintained control of the aircraft, departed the area, and recovered at Ubon without further incident. Two nights later, *The Arbitrator* commanded by Lt Col William "Bill" Schwehm was working a target in Laos when a salvo of 37mm AAA was fired. One round impacted and exploded on the underside of the fuselage approximately one foot behind the FLIR operator's seat. The booth was damaged and the operator's seat sustained several severe gashes on the backrest, but the sensor operator was saved by his armor-plated seat. The aircraft was recovered at Ubon Airfield without further incident but was out of commission for eleven days.[31]

On 9 March 1969, Lt Col Schwehm's crew was back in action on *The First Lady*, working an area in central Laos. Hampered by extremely poor weather, haze, and low clouds, the crew managed to sight thirty-five trucks and ultimately destroy twenty-one.

On 14 and 16 March 1969, Capt Mauri Galey's and Lt Col Ellis Von Haven's crews destroyed a total of thirty-five trucks, started twenty-nine fires, and set off fifteen secondary explosions in enemy storage areas. On 21 March, Maj Duane Phares dodged intense AAA fire, while IR operator Capt Larry Flinn sighted a tractor/trailer fuel or oil truck traveling southeast of Route 914. Phares began firing at the truck, but missed, causing its driver to speed up to about forty-five MPH.

> "Lt Col Simone was in the right seat and I was the IR. I located and tracked the truck and the first couple times we fired on it the driver sped up so our 20mm rounds kept hitting the road behind it. I adjusted the aim point a bit and we hit closer, but still behind it. Then I applied some aggressive lead and aimed well out in front of the truck and we finally hit it. It was a massive "blower" that burned for hours afterwards." **Lt Col (Ret.) Larry Flinn**

31 16 SOS Historical record, 1 Jan-31 Mar, 1969

While *Spectre* crews were getting better, so were PAVN gunners. On 27 March 1969, Capt Galey's crew took off on *The First Lady* (call sign *Spectre* 01) for a routine truck hunting mission near Route 914 in Laos. After destroying one truck, they encountered a proficient 23mm AAA gunner who convinced them to move eastward, out of range. After the IR located two semi-tractor/trailer trucks traveling close together, Galey fired a short burst, missing both. The trucks began to accelerate with the gunship crew in hot pursuit. About the time they were in proper geometry to fire, a 23mm projectile exploded on the forward inside edge of their right elevator. Undeterred, Capt Galey let loose a long burst of 20mm, hitting the second truck, causing it to explode. The crew was unaware their right elevator spar and hinge was damaged much more seriously than thought, but they successfully recovered at Ubon. After landing, the elevator literally fell off while crew chiefs began repairs; undoubtedly, it was a close call. Capt Galey's copilot on this mission was Maj Charlie Spicka, who had been hit twice on previous missions. Naturally, after being hit three times, fellow crewmen named Maj Spicka "Old Magnet Ass." A running joke in the squadron was Charlie could not get a crew together anymore since they would surely be hit, but it was only a joke and crews continued to fly with him, nevertheless.

By April 1969, crews were taking battle damage more frequently and the general consensus was risk increases in bright moon and star conditions when backlit by a thin overcast condition.

> "It should be noted that all recent AAA battle damage incidents occurred when the aircraft was flying in conditions of a sizable moon. By virtue of its physical size, slow airspeed and required low operating altitude the AC-130 proved to be a lucrative target for enemy gunners when it was flown in a light sky environment. Furthermore, the addition of a high thin overcast usually silhouettes the aircraft against the cloud layer. On missions such as those, the aircrew is able to see the aircraft shadow on the ground. Therefore, AAA was always extremely accurate and degrading to the gunship mission accomplishment when operating under those conditions." 16 SOS Historical record, 1 Apr-30 Jun, 1969. (U)

After a string of battle damage incidents, squadron leadership published directive HOT POOP Item Number 168, laying out considerations to decrease vulnerabilities to PAVN guns. Crews were directed to avoid highlighting and remain unpredictable as much as possible. There was also concern the PAVN was monitoring *Spectre* radio transmissions to improve the probability of shooting down a gunship.[32] To mitigate the issue, the 7th Air Force commander directed 16 SOS to use rotational call signs of the 8th TFW

32 Intelligence collection after the end of the war implies the PAVN was indeed monitoring AC-130 radio traffic.

instead of the *Spectre* call sign. Crews felt the change caused needless confusion when working with ground parties, but the directive went into effect nevertheless.

During the month of April 1969, the PAVN pushed hard to move massive quantities of supplies and equipment south before the end of the dry season. *Spectre* crews continued nightly missions in a varied AAA environment, searching for truck parks and traffic. On the night of 3 April, Lt Col Von Haven's crew (call sign Combine) launched on *Mors De Caelis* to work the heavily defended Routes 914B and C plus 92C, D, E in the central-eastern portion of Laos. After destroying three known storage sites in a lightly defended area, the crew awaited the arrival of their F-4 escorts before proceeding into the more heavily defended areas.[33] After firing on several more storage sites, the crew was searching along southern Route 92 when 37mm AAA fire struck the front of aircraft in the nose wheel well, directly under the flight deck, wounding the flight engineer MSgt Lopez and NOD operator Capt Lotshaw.

> "The nose gear assembly, engine power, nose gear steering and rudder controls were damaged and two crewmembers, the flight engineer and the NOD operators received injuries as a result of the blast and shrapnel which penetrated the cockpit and crew entrance areas." 16 SOS Historical record, 1 Apr-30 Jun, 1969. (U)

33 Unit Historical Records, 16 SOS 1 Apr-30 Jun, 1969

As the crew assessed damage level, Lt Col Von Haven discovered the rudder was completely useless, number three engine was uncontrollable, number two engine instruments were inoperative, and all utility hydraulic pressure was lost. Fortunately, Lt Col Von Haven took rapid action to get the aircraft under control and exit the area towards the nearest landing field. While en route, the number two engine was shut down, while the FE, bleeding from a wounded left leg, inspected the nose landing gear. Finding both tires blown out, a throttle cable cut, hydraulic lines severed, and extensive damage to the gear doors, he reported the damage to the aircraft commander. By then, with one engine out and no rudder control or hydraulics, Lt Col Von Haven and his co-pilot, Lt Col Sherman could have easily lost control of the aircraft. The only directional control they had was manipulation of the outboard engines, and that's what it took to get the aircraft back to Ubon. On short final, the FE lowered the nose gear manually while the number three engine, stuck running at full power, was shut down. The crew landed without further problems, saving a valuable aircraft. Though the aircraft was a mess, it survived, and dedicated maintenance professionals repaired it within about a week.

On the night of 7 April 1969, four days after their close call on *Mors De Caelis*, Lt Col Von Haven's crew launched on *Gomer Grinder* (call sign Schlitz). Once in their working area, they searched for about thirty minutes and destroyed two vehicles under light AAA conditions. Schlitz was then diverted to one of the most heavily defended areas of Laos, Route 911, just south of the Mu Gia Pass. Searching the roads and waterways, they spotted twenty-three vehicles and began firing. As expected, PAVN gunners responded with heavy 37mm AAA fire, forcing Von Haven had to break out of orbit many times to avoid being hit. Lt Col Von Haven's crew, however, did not stop their attack until all trucks were burning on the roadway. Since they killed every truck they found, it was the first 100 percent kill rate mission for any AC-130 crew. Several nights later, Lt Col Schwehm's crew tied Lt Col Von Haven's record.

On 19 April 1969, Maj John Cahoon's crew (call sign Tampa) worked routes 23C between Delta Points 63 and 73 and Route 911 between Delta points 68 and 18. It was a moonless night and the gunship was all but invisible at 5,500. PAVN AAA was moderate to heavy during their search and most AAA was inaccurate. When a convoy was detected, Maj Cahoon rolled into firing orbit and attacked with 20mm and 7.62mm guns. Ultimately, Cahoon's crew destroyed thirteen trucks and damaged two while an F-4 escort (call sign Hornet) working with Tampa, destroyed seven more trucks and one 37mm AAA site.[34]

By the end of April 1969, AC-130 crews had attained a high skill level in both offensive and defensive tactics. PAVN truck drivers and AAA gunners feared the AC-130, which they called "the Thug," and it was common for drivers to abandon their trucks whenever an AC-130 arrived overhead. Sometimes drivers responded by driving wildly and erratically to make a more difficult target, but gunship crews learned to be patient when tracking them. One night,

34 From official report of Maj Cahoon's mission #0546/7

a particularly confused driver was tracked while sensor operators tried to figure out what he was doing from one moment to the next. When a crewman quipped, the driver acted like 1960s television comedy character Gomer Pyle, the title stuck. Ultimately, the name "Gomer" was popularized in the squadron and was used to describe all PAVN personnel on the Ho Chi Minh Trail.

Near the end of the 1969 dry season, AC-130 crews had racked up a spectacular number of trucks damaged or destroyed, gaining the respect of 8th TFW fighter pilots and the 7th Air Force chain of command. In May, the 16 SOS received a congratulatory message from the 7th Air Force Headquarters for destroying over 29 percent of all trucks in Laos while flying less than 4 percent of total truck hunting sorties.

> OO RUSDIK
> DEE RESVB 1784 1280430
> ZNY SSSSS ZOK JPCCO
> P 090425Z MAY 69
> FM HHQ 7AF TAN SON NHUT AB RVN
> TO RUSDIK/8th TFW UBON RTAFT THAI
> INFO RUSDT/7/13AF
> RUMMWKA/13AF
> BT
> SUBJ: SPECTRE OPERATIONS (U)
>
> 1. THE CONTRIBUTION OF THE AC-130 GUNSHIP TO THE AIR INTERDICTION CAMPAIGN IN LAOS HAS BEEN TRULY

MAGNIFICENT. DURING THE PERIOD FROM JANUARY THROUGH APRIL THE SPECTRE ACCOUNTED FOR OVER 29 PERCENT OF THE TOTAL DESTROYED AND DAMAGED TRUCKS IN LAOS WHILE FLYING LESS THAN 4 PERCENT OF THE TOTAL SORTIES USED TO ATTACK MOVING VEHICLES. THE RESULTS FOR APRIL ALONE WERE EVEN MORE IMPRESSIVE WITH THE SPECTRE ACCOUNTING FOR OVER 44 PERCENT OF THE DESTROYED/ DAMAGED TRUCKS WHILE FLYING ONLY 3.7 PERCENT OF THE SORTIES. THE 1400 PLUS TRUCKS DESTROYED OR DAMAGED BY THE SPECTRE DURING THE PAST FOUR MONTHS HAD HAD A SIGNIFICANT IMPACT ON THE ENEMY'S ABILITY TO SUSTAIN HIS FORCES IN SOUTH VIETNAM AND SOUTHERN LAOS AND WAS CERTAINLY A MAJOR FACTOR IN LIMITING THE MAGNITUDE OF THE NVA'S NORTHEAST MONSOON OFFENSIVE, HOWEVER. WITH THE FULL ONSET OF THE SOUTHWEST MONSOON SEASON LESS THAN A MONTH AWAY, WE CAN EXPECT THE ENEMY TO MAKE A CONCENTRATED EFFORT TO INFILTRATE ESSENTIAL MEN AND EQUIPMENT THROUGH LAOS INTO SVN BEFORE THE HEAVY RAINS BEGIN.

> TO COUNTER THE EXPECTED BUILDUP IN ENEMY INFILTRATION WE MUST BUNCH OUR MUSCLES AND FLY AS MANY PRIME TRUCK KILLING SORTIES AS POSSIBLE DURING THE NEXT FEW WEEKS, THEREFORE, WE ARE SCHEDULING FIVE AC-130 SORTIES PER NIGHT FOR THE NEXT SEVERAL WEEKS OR UNTIL THE WEATHER PREVENTS THEIR EFFECTIVE EMPLOYMENT, AT WHICH TIME WE HOPE THE RAINS WILL HAVE ADDED TO OUR INTERDICTION CAMPAIGN BY SEVERELY LIMITING THE ENEMY'S TRAFFIC FLOWING THROUGH LAOS. GOOD HUNTING.

GP4

BT 1784

Though the results were phenomenal, they were not easy to achieve. Crews routinely dealt with equipment failures, especially the fire-control system and IR set. Several times, crews sighted multiple vehicles, fired, and either missed or achieved sporadic hits and minimal damage. The situation was extremely frustrating for crews, especially when they took heavy AAA fire without the ability to inflict damage to the enemy, but there were no options other than to deal with it.

Nobody could have predicted the AC-130's historical first that occurred on the night of 8 May 1969. Lt Col Schwehm's crew (call sign Bennett) took off on *The Arbitrator* to work along routes 110 and 96E in the extreme southeast portion of Laos. Their mission progressed normally and enemy AAA was light and inaccurate. Bennett spotted nineteen trucks and had destroyed twelve before the NOD operator sighted what he believed was a helicopter flying about one thousand feet above the jungle. PAVN helicopters had been on the target list for months, but it was rare to see one, let alone get a shot at one. The NOD operator continued to track the ghostly image until it landed in a clearing and virtually disappeared into the jungle. It took more than twenty minutes for *Moonbeam* to clear Bennett to fire. About the same time the crew was cleared, Lt Col Schwehm thought he saw a dark image that appeared to be a helicopter lifting off. He immediately opened fire with the 20mm guns, covering the area. There were no secondary explosions or fires detected, but he believed he had shot down the helicopter. During debrief, he reported the suspected kill. During the following days, the wreckage of a helicopter could not be located, so Lt Col Schwehm was not given credit for an air-to-air kill. Nevertheless, the story took on a life of its own resulting in the slogan "The Fabulous Four Engine Fighter" being emblazoned on the front of the 16 SOS operations building.

Commando Hunt ended on 30 April 1969 just as the annual rainy season began. It was a hard six months for *Spectre* but morale and confidence were high while crews and maintainers developed a regular routine. Facilities and equipment were improving, a nose dock was completed for phase inspections, and experiments were underway to better capture AC-130 BDA. A new AAD-4 FLIR equipped with a video recorder arrived with aircraft 54-1623 (*Ghost Rider*) which provided much better imagery than the original FLIR. A fixed NOD, bore-sighted to the guns, was mounted in the upper half of the left paratroop door opening to better film BDA results.

On 24 May 1969, the 16th SOS incurred its first combat loss. That night, Lt Col Schwehm commanded *The Arbitrator* (call sign Carter) on a mission into the heavily defended routes in central-eastern Laos. When working in the vicinity of the road crossing of Routes 920 and 92C, Sgt Jack Troglen, the Illuminator Operator, called out "Triple-A at 6 o'clock; going to be accurate!" Seconds later, a number of 37mm shells exploded on the underside of the right wing and tail. Lt Col Schwehm turned the aircraft toward Ubon. Shortly afterward, both hydraulic systems depleted because of severed lines, and the aircraft began to roll right and descend. The pilot's flight controls were useless without hydraulic power, but all four engines were operating normally. Lt Col Schwehm knew the only thing keeping the *Arbitrator* from falling from the sky was airspeed and a "nose-on" condition. Fortunately, Schwehm was able to regain basic directional control with differential power and aileron trim.

> "The aircraft started to roll to the right and started a descent. I had no elevator, rudder or aileron control. Using differential power

> and aileron trim, we were able to right the aircraft. Rudder and elevator trim were inoperative. I asked for a crew check and was told the IO was injured." Statement by Lt Col Schwehm, 16 SOS Historical record, 1 Apr-30 Jun, 1969.

Events in the back of the aircraft were dire. A 37mm projectile had exploded on the fuselage near Sgt Troglen's position, seriously wounding him and striking photographer Sgt Edward Marerro. Sgt Marrero advised the crew Sgt Troglen was losing a lot of blood and had a weak pulse. Crewmen carried Sgt Troglen into the sensor booth to better assess his wounds and administer first aid.

> "I heard approximately four rounds going off in the 6 O'clock position. The rounds I feel struck or were striking the aircraft. They appeared to have a dull thud type sound rather than the sharp crack sound when the rounds pass close by. I turned in that direction to see where they were impacting and experienced a slamming sensation in my left thigh. I looked in the direction of the IO who was about 15 feet further aft. I noticed he was limply hanging over the ramp secured by the restraining cable which I grabbed at the time. I pulled him back on the ramp which was inclined at a 30-45 degree angle. I

> turned him over on his back and he appeared to have severe head wounds as well as a wound in his buttocks." Statement by Sgt Edward Marerro, 16 SOS Historical Record, 1 Apr-30 Jun, 1969.

While Lt Col Schwehm and his copilot Maj Jerry Piehl carefully manipulated throttles and aileron trim, Flight Engineer SSgt Cecil Taylor rushed to the back of the aircraft to close the cargo door to provide the pilots better control. While there, he saw extensive fragment damage on the rudder boost hydraulic system. He also noted both hydraulic systems were depleted and the elevator cables were cut. At that point, everyone realized the aircraft could easily depart controlled flight. Airspeed was critical since the pilots could only maintain level flight when dangerously close to stall speed since adding power resulted in a climb. To change the center of gravity and thus the ability to add power, Lt Col Schwehm ordered all crewmen to the front of the aircraft to add weight to keep the nose down. From that point on, he was able to add power without climbing. In an amazing feat of airmanship, the pilots flew the aircraft back to Ubon. Finally, with the runway in sight, Lt Col Schwehm was advised Sgt Troglen had died from his wounds. He then ordered everyone except for Maj Peihl and SSgt Taylor to bail out.

> "I would estimate that we were airborne half an hour at least from the time we got hit until we began bailing out. I think I was the

> second or third man out. The man in front of me hesitated in the door, but Ken Stockard's boot in the backside quickly ended that. I didn't want any assistance, so I just stepped up and out. The time between getting hit and bailing out was more than enough to re-tighten the straps of my harness several times and to be thoroughly saturated with adrenaline. One of the more consequential mistakes I made was forgetting to count to three after exiting the aircraft. I was still in the prop wash and probably traveling about 160 knots when I pulled the D-ring on my chest pack. The opening shock was severe, resulting in some separated ribs and some very sore neck muscles from the whiplash but the parachute performed as advertised."
> **Lt Col (Ret.) Norman Clanton,** Table Nav.

Knowing he would only have one attempt to land, Lt Col Schwehm was on a short timeline. The runway was in sight as he manipulated throttles and trim to descend while SSgt Taylor manually lowered the main landing gear. Arriving over the runway threshold, Lt Col Schwehm pulled back power until the aircraft hit hard on the runway, nose gear first. The impact jolted the *Arbitrator*, which began to veer off the runway to the right. Unfortunately, unbeknownst to Lt Col Schwehm, the runway had been foamed, which exacerbated control problems. As the aircraft careened out of control, pilots tried to reverse engines to no avail. When Lt Col Schwehm applied power to the right engines to bring the aircraft back onto the runway, the aircraft was unresponsive as it collided with a mound of dirt and barrier shack, shearing off the right wing.

> "With the use of aileron trim and power we managed to get the airplane aligned with the runway. When power was reduced for touch down, the aircraft nosed down and hit the runway. I reversed all four engines but the aircraft started off the right side of the runway. It was apparent that the engines did not reverse so I applied power to #3 and #4 engines with the right wing. We went over a large mound of earth and hit BAK-13 shack 12 barrier shack." Statement from Lt Col William Schwehm, 16 SOS Historical record, 1 Apr-30 Jun, 1969.

When the aircraft hit a second barrier shack, it came to a stop. Both pilots quickly jumped off the flight deck into the cargo compartment for emergency egress. Once outside, they discovered the NOD operator, Maj "Monty" Montemayor did not bail out and had egressed after the crash, but there was no sign of SSgt Taylor and the aircraft was engulfed in flames. Since firefighters had depleted firefighting foam on the runway, none was left to fight the fire. Water trucks arrived but had little effect as the aircraft burned out of control to total destruction. While fighting the fire, a fire truck stalled and was subsequently destroyed. The crew members who bailed out were rescued by ARRS helicopters that night and early the next morning.

Nearly the entire squadron watched helplessly throughout the night as *Arbitrator* burned on the runway. The spectacular feat of airmanship demonstrated by Lt Col Schwehm and Maj Piehl was overshadowed by the shock and death of SSgt Taylor and Sgt Troglen. Aftershocks included prevalent and forceful theories and accusations. Crewmen suspected *The Arbitrator* was hit because it was visible to PAVN AAA gunners under the illumination of a full moon, but defensive tactics were questioned as well. Nevertheless, *Spectre* crews were back in the air two days later hunting trucks on the trail.

The wreckage of aircraft 54-1629 on Ubon Airfield, May 1969. (USAF Photo)

Lt Col Schwehm's crew after the loss of *The Arbitrator* which took the lives of SSgt Taylor and Sgt Troglen on the night of 24 May 1969. (USAF Photo)

On 29 May, Maj Duane Phares' crew (call sign Carter) boarded *Gomer Grinder* for a mission to Routes 23B&D, 125B, 91A, and 237. These Routes ran through areas in Laos that were considered to be fairly permissive with only light AAA potential. After joining with their F-4 escorts, Waco Flight, over Route 91A, they discovered the area was unworkable and requested *Alley Cat* reassign them to another search area. Weather was a problem there also, so Carter was redirected to Route 911 near Delta Point 18, where they found a moderately workable hole in the weather. Upon arrival, Maj James Staten, the NOD operator, located one truck and Maj Phares fired on it, but broken clouds at three thousand ft. hindered target acquisition. Capt Schiuten, the table Nav, asked *Alley Cat* for another working area. Since a C-130, *Blind Bat,* had located six trucks, Maj Phares was directed to Route 911, Delta 18, unaware they were about to fly directly into an AAA trap. All this time with Blind Bat 01 overhead, PAVN gunners held their fire, hoping to lure in a gunship. As Carter arrived over the area, Blind Bat 01 cleared off and Carter went to work. They located three trucks but lost them in the undercast conditions. While breaking in and out of cloud cover, Maj Phares was just about to fire when several PAVN gun sites opened fire simultaneously.

> "While trying to locate the trucks again, we were fired on by two AAA sites, one at nine o' clock and one at two o'clock, both accurate and both fired ten rounds. The site at two o'clock had a nine-level gunner and we spurred a hit in the left wing leading edge midway between number one and number two engines. After the hit, number one throttle was at the 90-degree position and engine power was at flight idle indication. Initially, it was extremely difficult to make a right turn to proceed west out the target area, I ordered a shut down of the number one engine then used full right rudder trim and full right wing aileron trim, by adding additional right rudder was able to make a very slow right turn." Extract from official statement by Maj Duane Phares, 16 SOS historical records, 1969.

Of the rounds fired at Carter, one hit the aircraft in the left wing, centered between the engines. A large portion of the leading edge of the wing had been blown off, causing drag, shudder, and vibration. Maj Phares and copilot Capt Louis Bartrand retained control of the aircraft as 37mm tracers streaked past the gunship, some missing by only a few feet. Flight engineer TSgt Richard Mahaffey scanned the number one engine, then shut it down for an oil leak. The aircraft was in trouble and even while set at full-take-off power, drag from the missing section of the leading edge and the loss of an engine resulted in a maximum-indicated airspeed of 135 knots. *Gomer Grinder* was descending at a rate of 100 feet per minute and the left aileron was inoperable. Mahaffey scanned the wing then returned to his seat to monitor engine conditions. By this time, Schuiten identified Nakhon Phanom as the nearest US base and passed its heading to the pilot.

> "I had no indication of fire in the "T" handle. I scanned the engine from the flight deck and no fire was present on visual inspection. I could not determine the condition of the engine at this time. I came back in the cockpit to monitor instruments again. Approximately 13 seconds later a low oil quantity light came on. I noted that #1 engine only had 3 gallons of oil. I recommend to the pilot to shut the engine down, which he immediately did. The prop went to full feather and shut down, no problem." Extract from official statement by

TSgt Richard Mahaffey, 16 SOS historical records, 1969.

In the rear of the stricken aircraft, IO SSgt Vincent Marrazzo and gunner Sgt Thomas Griebel completed a damage assessment while the rest of the gun crew began to jettison ammunition, logbooks, brass, and even their toolboxes to lighten the aircraft. After declaring an in-flight emergency, the crew safely landed, then evacuated the aircraft while it was still on the runway. Once *Gomer Grinder* cooled down, crew chiefs were cleared on to inspect the damage. Aside from extensive damage to the leading edge of the left wing, the airplane was also leaking fuel and hydraulic fluid.

This event ended well but could have been catastrophic. Fortunately, the crew was highly experienced and well disciplined. They handled a potentially fatal situation as if it were just another routine mission on the trail. The aircraft remained at NKP for two weeks to be repaired by an AFLC RAM team specializing in battle damage repair. The aircraft was flown back to Ubon on 16 June by Maj Cahoon and Maj Ward Meyer.

By June 1969, the weather had taken a turn, causing heavy rainfall and much less traffic on Ho Chi Minh Trail. Standing pools of water and thick mud prevented nearly all heavy traffic. Crews had difficulty locating targets, even with the IR sensor. With little truck activity on the trail, AC-130 crews were tasked to support ground forces more often. In late June, the squadron was called upon four times to assist in the defense of the neutralist-held camp at Muong Soui, located forty-seven miles east-southeast of Luang Prabang. The camp, labeled Lima Site 108, was a small listening post located on high jungle terrain in a remote part of Laos. On 25 June, the camp came under heavy mortar and rocket fire from advancing North Vietnamese troops. *Mors De Caelis*, commanded by Capt Louis Bartrand (call sign Altoona) was searching an area in *Barrel Roll* when they heard an urgent call for help from Lima Site 108. *Altoona* established radio contact with the camp controller (call sign City Hall), who directed Capt Bartrand to fire on enemy troop positions surrounding the camp. Once Altoona began firing 20mm and 7.62mm, the enemy began to take heavy casualties. As seen before, multiple waves of attackers attempted to breach the camp, but all were met with *Spectre* gunfire. During Altoona's two-and-one-half hours on station, numerous aggressors were killed and sixteen secondary explosions were detected. All was quiet when *Mors De Caelis* headed for home.

The following evening, Maj Piehls' crew (call sign Buick) took off on *The First Lady* and was immediately directed by Blue Chip to report to Site 108 to repel another attack. When they arrived overhead, Spooky 03 was already working the target, so they remained in holding. Once called in, they were directed to search for enemy positions and tanks that were reported to be heading for the camp. Though Buick was hampered by weather, they stayed on station searching for targets until low on fuel. After landing at Udorn for fuel, the weather closed in completely and crewmen found a hydraulic leak that grounded the aircraft. Obviously, *The*

First Lady spent that night at Udorn. On 29 and 30 June, two more missions were flown to Lima site 108 with limited success because of bad weather conditions.

Viet Cong guerillas welcomed bad weather since they operated with near impunity when aircraft were unable to fly. Nevertheless, AC-130 crews remained on alert hoping to find a break in the weather good enough to provide fire support. Many TIC requests ended without the crew ever seeing the ground, or, in some cases, ever taking off. On 8 July, Capt Jack Starr's crew (call sign Tampa) took off on *Ghost Rider*, then proceeded to their assigned area in *Barrel Roll* to search for truck traffic. Unable to see the ground because of low clouds, they shifted to a secondary area which was also obscured. After checking in with a ground FAC, they were able to spot a red flare and white strobe marking the location of the ground commander (call sign Hunter), who advised the gunship crew he was under attack. To make sure they had the friendly position identified, Tampa used their overt illuminator to verify Hunter's position. Hunter then provided range and bearing of the enemy position about one kilometer from the camp. Lt Col Simone suggested the crew use the fire control computer to establish offset-firing parameters while sensor operators continued to track the friendly position. The Table Nav, Maj William Crafton III, entered an offset of 999 meters and a heading of 091 degrees into the fire-control computer. He then advised Capt Starr he could only fire on a heading between 270 and 030 degrees. While the NOD operator Capt Gerald Harris tracked the friendly position, Capt Starr prepared to fire using offset guidance, requesting #2 and #3 20mm and #3 and #4 7.62mm guns be armed. When all firing parameters were met, Capt Starr fired a one-second burst to ensure all was in order before commencing an aggressive attack. What should have been a successful offset attack was anything but successful.

> "TO on time cleared to work fragged area, however, unable to work roads because of weather. Further cleared to work with Hunter. Identified his position and he cleared us to fire on enemy position 1 kilometer to the East. I sent in an offset of 999 meters/091 degrees. The NOD tracked a strobe in the camp, we rolled in and when everything looked normal, we fired. When bullets impacted the ground FAC very excitedly told us we hit his position. NOD had perimeter in sight and he says all flashes were outside the camp. We informed Alley Cat and Tuna 68. Tuna 68 said he would go in at first light and investigate." AC-130 mission report, MSN NR 1529, 8 July, 1969.

To everyone's surprise, the burst impacted close to Hunter's position. It was a close call but thankfully, nobody was killed. The fact that Capt Starr elected to shoot a short test burst undoubtedly prevented a catastrophic loss of life. Lt Col Simone, the 16 SOS Operations Officer was on board during the event and immediately prohibited all offset-firing until the issue could be resolved. The situation was thoroughly investigated by ASD via live-fire testing at Lockbourne. Problems were discovered with the fire-control computer and training. Eventually, the problems were resolved, but offset-firing procedures remained a touchy subject for some time.

On 29 July 1969, Lt Col Thomas Simone assumed command of the 16 SOS. At the same time, some of the initial cadre reached the end of their one-year tour and began to depart for other duty stations. Not surprisingly, *Spectre* combat tours always ended with wild parties, debauchery and fanfare. Of course, alcohol played a role in the festivities, right down to naming a crewman's last mission a "champagne flight."

> "When a crew member flies his "Champagne Flight" (last mission before rotating back to "the world"), he was hosed by a firetruck, placed on a party wagon which delivered him to the Airman's or NCO, or O Club, and was handed a bottle of champagne. Somehow, I got 2 bottles." **Sgt Len Johnson, Gunner,** 1969.

End-of-tour parties were held in high esteem in the 16 SOS and were a reason to celebrate. Formal attire for an event began as USAF issue K2B Flight Suits dyed black, but eventually crewmen had custom "Party Suits" made by a tailor shop downtown. Crewmen were rightfully proud of their party suits which were adorned with patches, pins, and emblems associated with the squadron. Enlisted wore black, and officers wore blue.[35]

End-of-tour changes resulted in crewmen being replaced on a one-by-one basis. The squadron was still short on gunner manpower since only three per crew were authorized, but four were needed. A continual battle with Air Force Manning and Personnel Center ensued which eventually resolved the issue.

By August 1969, extremely poor weather conditions in Southern and Central Laos caused the cancellation or curtailment of most missions. Many roads on the Ho Chi Minh Trail were unpassable, but some level of supplies continued to flow to the south via bicycle porters and indentured servants on footpaths. Farther north in Laos, the weather and road conditions were somewhat better, but not significantly. During the decrease in mission activity, aircraft were ferried to CONUS for modification and avionics improvements. *Plain Jane* aircraft returned in the fall, ready for combat with a 200-channel head for the AAD-4 FLIR, Dual Axis Gyro, and TRIM 7 electronic countermeasures device.

By October 1969, the weather had improved enough to permit interdiction missions on the Ho Chi Minh Trail, mostly in the *Steel Tiger East* and *Barrel Roll* areas of Laos. Since there were many crewmen who had recently arrived for their tour, the squadron initiated an internal Combat Crew Training program to increase the effectiveness of low-time and inexperienced crew members. In effect, new crews

35 Any crewmen who survived being shot down wore a maroon party suit.

were always sent into light AAA areas at the end of the rainy season before being thrust directly into heavy combat when regular dry season interdiction missions resumed.

Operation Commando Hunt III began on 1 November 1969. The roads on the trail were almost completely dry and trucks were moving in great numbers. On 6 November 1969, Maj Mackinnon's crew flew *Gomer Grinder* (call sign Nettle) through the 911 trail network, finding four trucks and destroying two. Continuing their search on Route 91, they discovered twenty-five vehicles moving southeast, but were directed by *Moonbeam* to leave the area since it had a flight of fighters inbound. Proceeding south, they discovered twenty more vehicles on Routes 239 and 23. Maj MacKinnon's crew proceeded to destroy ten trucks before running out of ammunition. Three days later, Maj Ward Meyer's crew flew *The Exterminator* (call sign Ringo) into an area near the 91A network and immediately detected a convoy of twenty-five trucks on a narrow trail. After firing on the lead truck, causing a fire, the remainder of the convoy stalled, unable to maneuver around the burning truck. Using proven AC-130 tactics, Maj Meyer destroyed the trail vehicle, trapping the entire convoy in between. Surprisingly, there was no AAA being fired, so the crew systematically fired on the convoy until they ran out of ammunition, destroying at least nineteen trucks. Maj Meyer's crew later described the scene below as a "burning, exploding holocaust in the jungle below."

Learning a lesson from the previous dry season, PAVN gunners were ready and waiting for the AC-130 with even more guns on the trail than the previous year. Even when escorted by F-4s, the AAA threat was a significant problem to deal with. The F-4s were vulnerable when attacking targets also. On 11 November, an F-4 flown by Capt Robert Tucci and Maj James Dennany was either hit by AAA or crashed while attacking a convoy while supporting an AC-130. As a result of increased numbers of 37mm guns on the trail, AC-130 crews found themselves spending more time dodging AAA than firing on trucks. It was also evident that PAVN gunners were aware the AC-130 usually flew at 5,500 ft. and had nearly perfected their engagement tactics whenever a gunship came overhead. The Gunship Program Office realized 5,500 ft. altitude had become a major survivability issue. It was painfully evident the *Plain Jane* model had reached its performance ceiling and there wasn't much more that could be expected from it.

Fortunately, as early as mid-1968, the Gunship Program Office predicted PAVN tactics would eventually negate AC-130 capabilities on heavily defended areas of the Ho Chi Minh Trail. Frustrated by LTV's disinterest in building an advanced AC-130, ASD sought approval and received funding to build an eighth AC-130 themselves. The program, titled *Coronet Surprise*, was supported by TAC, PACAF, commercial enterprises and the USAF Academy. Many of the original gunship pioneers, including Maj Ron Terry, were deeply involved in the project. C-130A, tail number 56-0490 was assigned to the project. Officially known as the *Surprise Package Aircraft* (SPA) and appropriately named *Thor*, it was built along the lines of a standard production model, but with eight special subsystems.[36] Armament changes consisted of replacement of all four aft guns with two 40mm guns that were capable of being fired from an altitude as high as 10,000 feet. An improved navigation/targeting system was added for improved targeting and a 2KW illuminator replaced the rarely used and cumbersome 40KW light. A Black Crow electronic emission sensor was added to detect running gasoline engines. A low light level television sensor with a laser target designator replaced the NOD. Other improvements such as a moving target indicator, digital fire-control computer, video recorder, and helmet sight were added. By the fall of 1969, initial testing at Wright-Patterson Air Force Base and Eglin Air Force Base, FL was complete and Tactical Air Command (TAC) directed *Surprise Package* to deploy to Ubon for combat trials.

> "Considering the introduction of new and unique aircraft, the supply problems associated with an unusual system, and the personnel problem due to temporary shortages of specialized mechanics, overall maintenance performance fared quite well. No small amount of credit for the tenacity and morale of the section was due to Capt Dave Wilson and his capable cohorts, Capts Mike Busby, Dick Coker, Bab Grant and Lieutenant Terry Thompson." 16 SOS Unit History, Oct-Dec, 1969. (U)

36 Thor was named "Super Chicken" by 16 SOS crews.

By the middle of December 1969, PAVN AAA guns were so prolific in AC-130 hunting areas that Capt Stebbins' crew (call sign Cobalt) took 587 rounds of AAA, with 297 considered accurate. Though the crew sighted at least ten vehicles, they were constantly driven off target by AAA. Even after rolling out and returning from different headings, and use of partial orbits, they continued to be driven off by intense AAA. After about a dozen attempts, Cobalt gave up and returned to base. It was evident something more than a change of tactics was needed to remain effective against PAVN trucks.

Surprise Package began flying missions on 12 December 1969 to test combat capabilities at altitudes as high as 10,000 ft, above the effective range of PAVN 37mm guns. By the end of December 1969, *Thor* was flying regular missions into some of the most defended areas of *Steel Tiger East* and *Barrel Roll* with near impunity. On 30 December, Maj Gentry's crew (call sign Tampa) boarded *Thor* with Maj General Robert Petit onboard as an observer. Their destination was a chokepoint on the Ho Chi Minh Trail known for heavy truck traffic. Using *Surprise Package's* new target locating systems, *Tampa* destroyed three trucks in short order with 40mm guns, then proceeded to fire on more trucks hidden under foliage. Determining they had discovered a large storage area; Tampa used their Loran target designation system to guide F-4 fighters onto the target. Within ten minutes, more than 100 fires had erupted over a 1,000 to 1,500-meter area that produced a column of smoke about 5,000 ft. high. Moving to the northeast,

Tampa detected an uncharted temporary bridge and fired on it as a vehicle was about to cross. Although AAA was heavy, *Thor's* higher operational altitude increase negated much of the 37mm threat and was clearly a step in the right direction of AC-130 capabilities. By the end of December 1969, seven AC-130A aircraft were flying out of Ubon, hunting trucks on a nightly basis.[37]

37 Aircraft numbers were 53-3129, 54-1623, 54-1625, 54-1627, 54-1628, 54-1630 and 56-0490.

CHAPTER TWO

VIETNAM, LAOS, AND CAMBODIA OPERATIONS, 1970–1973

Aircraft 54-1630 *Mors De Caelis* was renamed *Azrael, Angel of Death* after conversion to Pave Pronto configuration. (USAF Photo)

By 1 Jan 1970, Commando Hunt III was in its second month of operation. The trails were dry and truck traffic was active. Every night, *Spectre* crews hunted for *Group 559* trucks within *Steel Tiger East*, *Barrel Roll*, and *Tiger Hound* VR areas. Morale was high and the atmosphere competitive as crews rivaled to earn the honor of being the named "crew of the month" by turning in the highest number of trucks destroyed/damaged. Though AAA was at least twice as heavy as it was during the previous dry season, there were vulnerable trail sections between gun sites that were prime truck hunting areas. *Spectre* navigators were key to ensure their crews remained beyond the range of AAA. At the same time, PAVN AAA gunners continually modified their tactics trying to lure an AC-130 into range. If a gunship crew was engaged by AAA from a previously unknown site, fighter escorts dropped cluster bombs on the site while *Spectre* continued to engage truck convoys. Crews were racking up impressive truck kill numbers and the Surprise Package aircraft (Thor) was setting one record after another.

In late January, Secretary of the Air Force, Dr. Robert Seamans visited Ubon to view AC-130 operations and the *Surprise Package Aircraft,* in particular. Impressed with *Thor*, and its record truck kills, he directed the aircraft to remain at Ubon after combat trials completed instead of returning to Wright Patterson AFB as planned. This decision was tactically sound, but *Surprise Package* avionics and fire-control differed greatly from *Plain Jane* models which added another wrinkle to the already stressed training environment.

Work was demanding. Aircrews maintained a rigid nighttime schedule, sometimes flying two four-hour sorties per night, landing only to refuel and rearm. Gunship maintainers worked around the clock preparing, launching, recovering, or repairing battle-damaged aircraft.

> "With eight aircraft on station, we planned to launch eight at 1900L and turn four at midnight. If seven were available, we turned

> three. The midnight was mostly limited by munitions load crews. However, we normally had at least one bird in battle damage repair. A repair team from Wright-Patterson were constantly at work." **Maj Gen (Ret.) Bob Patterson,** pilot.

Steel Tiger East was a busy area for air traffic. Every night, AC-130, AC-119K, B-57, and F-4 aircraft operated in a rather confined area in blackout conditions. To limit the probability of midair collision, *Spectre* aircraft became sector controllers for each crew's assigned VR area. AC-130 mission planners routinely varied VR sector assignments to prevent predictability. Since *Group 559* commanders would not know where and when an AC-130 would arrive overhead, listening posts were established at numerous points along the trail. This early warning system was nothing more than a soldier with a field phone who alerted his commander of the direction of any AC-130 passing overhead. Commanders then passed word for AAA crews to man their guns and lay in wait for an AC-130 to fly within range. *Group 559* leadership desperately wanted to learn more about the way an AC-130 worked since it attacked in a different way than any other aircraft. Since it was obvious the gunship always flew a geometric orbit when attacking, PAVN gunners often held fire until the AC-130 was within a predictable part of their orbit. Spectre crews countered by avoiding predictably as much as possible by the use of

S-turns, off-axis penetration of road segments, and random search patterns.

The first big mission of 1970 occurred on 15 January when Maj Gentry's crew (call sign Rufus) flew *Thor* to a rendezvous point in *Steel Tiger East*. After joining F-4 escorts, Rufus proceeded to their assigned VR sector and found it unworkable for weather. After receiving re-tasking by *Moonbeam*, they were sent to work Route 92E&F. Rolling in over the target area at 8,500 ft., they spotted a large number of trucks parked in an area; some appeared hot and others cold. Rufus opened fire while PAVN gunners returned fire. Maj Gentry's crew remained on station for about an hour and a half, ultimately destroying thirty trucks and damaging two, a new single mission truck-kill record for the 16 SOS. During their time over the target, they fired 2,500 rounds of 20mm and 351 rounds of 40mm, while taking 115 rounds of accurate AAA and 268 rounds of inaccurate AAA. The aircraft set a new standard for AC-130 performance and returned to base without incident.

Eleven days later, *Warlord* (call sign Gavel), commanded by Lt Col Brevig was engaged by 1,300 rounds of AAA in a two-hour period, setting a new squadron record for AAA received. During two hours over the target, Brevig continually broke away from accurate AAA, came back, fired, and was forced to break away time and time again. Nevertheless, Brevig's crew managed to destroy eighteen trucks and damage five. About halfway through their mission, while breaking out of orbit to avoid AAA, they received a SAM launch indication, but no launch was observed. The potential of a

SAM engagement convinced them to depart the area immediately and search elsewhere.

> "Cleared to working area after rendezvous with Cactus at 1845. At first target encountered intense AAA accurate, and SAM activity/launch light so departed area immediately. Found multiple targets in target #6 vicinity. RTB for fuel & ammo at 2200. Moonbeam cleared BAT 15 to 92B at 2115 denying us working there effectively. Called Wolfpack at 2220 with ETB 2245." Mission #1120 report, 16 SOS historical records, January-March, 1970.

Since the crew didn't see a launch, there was no positive confirmation their radar warning receiver detected actual SAM activity, or if the system was malfunctioning. Real or not, potential SAM activity in AC-130 working areas was a great concern. The SA-2 system was capable of shooting down a gunship, and crews had no defense against it other than maneuver and sheer luck. Since SA-2 activity could not be validated, missions continued to be scheduled as usual.

Another significant mission was flown on the night of 31 January 1970, by Maj Robert Kindred's crew, (call sign Mazy). After a routine take-off on *The Exterminator*, they joined F-4 escorts and proceeded to their assigned working area in *Steel Tiger* VR-4 near Tchepone, Laos. While searching, Mazy discovered a number of trucks moving down small trails off the main road. Maj Kindred's crew systematically fired on the 34 truck convoy, ultimately destroying 27 of them. During their mission, they successfully evaded 560 rounds of AAA, with 213 rounds accurate. Subsequent AC-130 missions were sent into the same area and upwards of 80 trucks were credited to AC-130 crews in a single week. Just after dark on Valentine's Day, Maj William Brooks, (call sign Usher) flew *Thor* into *Steel Tiger East* VR-4 near *Tchepone*. After finding a large assembly of trucks, Maj Brook's crew fired until they ran out of ammunition. Ultimately, *Thor* destroyed 43 trucks and damaged 2 while receiving 258 rounds of AAA. This mission surpassed all AC-130 single mission truck-kill records for the 16 SOS.

Because of increased traffic, AC-130 crews racked up spectacular truck kill numbers, but AAA was heavy. On the night of 18 February, *The First Lady* was battle damaged by AAA. The following night, *The Exterminator* and *Ghost Rider* were hit by AAA.[38] Furthermore, aircrews were concerned that high moon illumination and light overcast conditions were a factor since their aircraft were back-lit and visible by a light gray high overcast. Another survivability concern was flying during a full moon, a condition *Spectre* crews called "Gunners Moon." To assess these concerns, a series of tests were flown overhead Ubon at combat altitudes, during a full moon with high overcast. At the con-

38 On Feb 18, The First Lady was hit by 37mm aft of the left main gear well. On 19 Feb, The Exterminator and Ghost Rider sustained 37mm hits.

clusion of the test, evaluators found it difficult to visually acquire a gunship unless it was crossing directly through the moon. Though not everyone agreed with the findings, missions continued to be flown regardless of moon illumination or high overcast.

> "HHQ may have had their opinions on moon illumination, but I will tell you every Spectre crew member knew the moon phase as well as their own names. High overcast was akin to being a fly on a white ceiling, regardless of the aircraft paint- black, camouflage, or gray.... a highlight mission for me was a Friday 13th mission with a full moon." **Maj Gen (Ret.) Bob Patterson,** Pilot.

In response, *Plain Jane's* altitude was increased to 7500 ft., to incrementally counter the AAA threat. There were tactical impacts at 7,500 ft. however, the 20mm was on the edge of ballistic performance, and hit probabilities were lower. From that point on, *Plain Jane* needed to fire more ammunition than usual to achieve desired results.

Typical 37mm AAA damage to AC-130A . (USAF Photo)

By 21 February 1970, the 16 SOS reported its 5,000th truck destroyed or damaged, far surpassing all other types of aircraft however, not everyone shared the same level of enthusiasm.[39] From the beginning of AC-130 operations, U.S. Army General William Westmoreland thought reports of trucks damaged or destroyed by the AC-130 were unusually high. Original 1968 BDA criteria had three categories: damaged, destroyed, and results not observed (RNO). Simply put, if a truck was hit and burned, it was considered destroyed, if hit but no fire started, it was considered damaged. If fired upon, with no hits observed, it was considered RNO, which was only a notation that a truck existed at that

39 Development of Fixed-Wing Gunships, Ballard, P-130.

location at one time. When the *Surprise Package* combat evaluation occurred, ASD engineers convinced 7th Air Force staff that a single hit with a 40mm projectile would destroy a truck. Though that position was contested by some, it was accepted by the 7th Air Force under the condition that two sensor operators had to agree the target was hit before a claim it was destroyed could be made.[40] Unfortunately, ASD criteria did not account for many variables in truck design, cargo, and vulnerability, and 16mm AC-130 BDA films shot by the aft NOD camera were not detailed enough to determine incremental damage levels, but nevertheless, it was the accepted criteria.

On the other end of *Spectre's* guns, *Group 559* drivers were inventive and resourceful. It was common practice to begin clearing the trail as soon as the AC-130 departed. Trucks hit but still operational were simply restarted and driven away. Trucks that could not move under their own power were towed to the nearest Binh Tram for repair. Destroyed trucks were pushed off the trail and covered with limbs and foliage. Trails were usually cleared by sunrise, so by the time photo-reconnaissance aircraft arrived the next day, little evidence of an attack remained.

There was no limit to the surprises one could find on the trail. Periodically, a crew would fire on targets that proved to be something other than what they actually were. On several occasions, what appeared to be a convoy of trucks turned out to be a true-to-life elephant train, laden with PAVN supplies. Since elephants appeared to be large

40 Ibid, P-134.

hot spots on the IR sensor and were typically spaced like truck convoys, they were targeted as if they were vehicles. Another crew noticed an errant round well off the trail had lit a fire which continued to burn like a candle all night; they had severed a fuel pipeline. There was no shortage of rumors and wild stories relating to *Group 559* drivers. Some were earned, but some were unproven folklore. One of the most popular rumors claimed Laotian prisoners were forced to drive while chained to a truck's steering wheel.[41] Though the rumor was never validated, it may explain why some drivers were observed speeding and driving erratically instead of stopping and running away like they usually did. Though it's hard to tell how drivers made their decisions, it's easy to understand how being chased by high explosive projectiles would cause panic. In many cases, when a driver made a high-speed escape attempt, he either ran into a tree or rolled off the edge of a steep cliff…and yes, *Spectre* crews were credited with the kill. Trucks hauling explosives were always exciting when hit, but the most prized target were fuel tanker trucks crews called "Esso trucks." Busting open an Esso truck always resulted in a spectacular explosion and fireball. There were secondary effects also since thousands of gallons of burning gasoline would flow downhill, engulfing more trucks in a fiery quagmire of destruction. Though gunship crews reveled a good "blower," intense fires typically illuminated the gunship, increasing vulnerability to AAA fire.

41 Though there is no photographic proof, friendly Commando road watch teams in Laos reported finding drivers who appeared to be chained to steering wheels of destroyed supply trucks.

Group 559 relied heavily on tactical deception including the construction of elaborate fake convoys on main roads. Their intent was to lure crews onto "easy kills" while real convoys covertly traveled on back trails. Trail workers used felled trees, destroyed vehicles, and whatever junk they could find to create decoys that appeared to be serviceable trucks. These fake convoys were almost always placed near major intersections which were routinely targeted by *Spectre*. More enterprising trail workers lit small fires under junk trucks to spoof the IR sensor operator to believe it was a running vehicle. *Spectre* crews rarely fell for the decoys beyond firing a few rounds since they did not react to hits the same way an actual truck did. Another PAVN deception tactic was to assemble a sizeable fake convoy for the sole purpose of baiting aircraft into a AAA trap. *Spectre* crews were sometimes fooled by these fake convoys but learned quickly that a target that looked too good was most likely a AAA trap, and an area to avoid.

Gaining experience with every mission, *Spectre* and *Nite Owl* pilots became more aggressive. Sensor operators learned to watch for changes in driver behavior to exploit weaknesses in enemy tactics. Spectre crews continued to use the tried-and-true tactic of engaging the lead vehicle first, the trail vehicle second, but often tag-teamed their attack with their F-4 escorts who dropped conventional or cluster bombs on the convoy. One of the most aggressive F-4 pilots named "Animal," was known for fearless attacks on gun sites, many times diving as low as one thousand ft. to pull AAA away from *Spectre*. Sometimes he would turn on his navigation lights to lure hesitant AAA gunners into firing at him while his blacked-out wingman was in trail, ready to bomb the gun as soon as Animal buzzed the gun site and began to climb back to altitude.

> "Intelligence reports indicate that the enemy is constantly improving his tactics. For example, enemy gunners will fire at a gunship while anticipating a counterattack by the gunship's escort. When the escort "rolls in" on the gun, they are waiting for him. An observed enemy elaboration of the above tactic involves the use of several gun emplacements. A first gun fires on the orbiting gunship. As the F-4 begins his attack, another gun opens up on him hoping to distract him from his original target. When the fighter is fully committed, another battery will then fire either at him or the gunship knowing they are momentarily safe from attack." 16 SOS Unit Record, April 1970- June 1970. (U)

Not always did a gunship crew find trucks, sometimes trucks found them. On 24 February 1970, *Warlord* (call sign Xray) commanded by Richard Dyer, was sent into an area where AC-130s infrequently searched. After scanning a remote trail near a major road for some time, no activity was noted. About the time Maj Dyer's crew was ready to move to another area, sensor operators spotted a single truck that seemed unusual and out of place. Curious, X-Ray wanted to see where it would lead, so they held fire but tracked it for about ten minutes. When the truck pulled off the trail and disappeared into the jungle, the IR sensor operator noted suspicious circular trails that appeared to disappear into the jungle. As the IR operator continued to search the area, he detected hot spots of various intensity covered by jungle overgrowth. Convinced they had discovered a camouflaged truck park, Maj Dyer's crew fired on hot spots nearest the exits first to block egress via the circular trails, then opened up on targets trapped inside. In short order, Xray and their F-4 escorts systematically engaged every hot spot they could detect until they ran out of ammunition. During the attack, Xray took 150 rounds accurate and 165 rounds inaccurate AAA but received no hits. Maj Dyer's crew was credited for the destruction of twenty-seven trucks, thirty-three explosions, and twenty-nine fires. Total truck-kills for the AC-130/F-4 team were thirty-eight trucks destroyed and heavy damage and destruction of materials stored in the truck park. The mission was a success solely because of the crew's insight, experience, and discipline to hold fire to let the single-vehicle led them to what was the largest truck park yet to be found.

By March 1970, daily rain and overcast conditions marked the beginning of the annual rainy season, turning many trails in *Steel Tiger East* into a soft, muddy morass. *Group 559* pushed hard to get as many supplies as they could to the south before the roads became unpassable. Likewise, AC-130 crews surged to meet increased traffic. During this period of heavy activity, *Plain Jane* aircraft routinely accounted for twenty or more truck kills during a single mission. On 8 March, *The Warlord* (call sign Ohio), commanded by Maj Ronald Hight, destroyed twenty trucks in *Steel Tiger* East, and on 10 March, *Warlord* (call sign Mopup), commanded by Maj Kenneth Wilson, destroyed twenty trucks in the heavy AAA environment southeast of Tchepone, Laos known as the Iron Triangle.

On 4 April, Maj Kenneth Wilson's crew (call sign Gausy), set another squadron record. Arriving on station at 1930 hours and escorted by F-4s of Killer Flight, Gausy coordinated with Batcat, a targeting cell of the *Igloo White Project*. Batcat passed suspected vehicle locations found via signals detected by covert ground sensors previously dropped near the trail. Using target location information provided by Batcat, Wilson's crew rolled in overhead, but could not detect any targets through the jungle canopy. Nevertheless, Maj Wilson began firing on the areas identified by Batcat. Though it first appeared they were firing into the uninhabited jungle, small fires and smoke began to rise from the jungle. Encouraged by results, Gausy continued to fire until they ran out of ammunition. By the end of their mission, Maj Wilson's crew was credited with the destruction of nineteen trucks, five damaged, and thirteen more with results not observed (RNO).

On 7-8 April, the 16 SOS was alerted to support a ground party at Dak Seang, Vietnam, which was under attack. Maj William Brooks' crew flew *Thor* (call sign Wolfpack 54) to a holding area near the camp after a brief stop at Da Nang for refueling. Arriving at their holding area in undercast conditions, they couldn't see the ground. Area Controller Elliot 11 placed them in holding for about 50 minutes while fighters worked the target. When finally called in, Maj Brooks crew descended to 4,500 ft. to get below the weather. When cleared to fire, they teamed with Shadow 46, an AC-119 gunship, and began engaging personnel targets. After several hours, Wolfpack 54 landed at Da Nang to refuel, then returned to Dak Seang by 0730 to join other aircraft which were taking turns striking enemy positions. By 1030, Wolfpack 54 returned to Ubon to prepare for another mission that night.

On 12 April 1970, Maj Ronald Hight assumed command of the 16 SOS just as most trails in *Steel Tiger* East VR sectors became unpassable. From that point on, traffic was light, and *Spectre* crews battled low cloud decks and nearly constant rain. While heavy weather and light traffic in Laos limited truck hunting missions, AC-130 crews spent time on ground alert to support friendly ground forces. At least some missions were flown into Vietnam, but the primary focus was the support of Laotian ground forces. By late June, constant wet weather not only hindered target acquisition, but the muddy surface was so soft that 20mm HEI fuzes would not function on impact, negating a crew's ability to see shot impacts and correct fire.

On the night of 21 April 1970, the weather in southeastern Laos was unusually good with two to five miles visibility and a light overcast. Maj Brooks' crew took off on *Warlord* (call sign Adlib 11) to search Route 96, a major trail through Laos. Adlib 11 arrived in their search area under a full moon at 7,500 ft. escorted by two F-4s (call sign Killer 01 and Killer 02). After establishing a firing orbit, Adlib 11 was engaged with fifteen rounds of accurate 37mm AAA. Maj Brooks quickly broke away from the AAA while Killer 01, commanded by Maj Richard Merkell, acquired the gun site, dove towards it, and dropped one CBU-24 cluster bomb. Believing the gun site was suppressed, Adlib returned overhead as ten inaccurate 37mm rounds passed to their rear. Adlib cleared Killer 01 onto the gun site once again while remaining in orbit. Just as the fighter began its bombing run, Adlib was hit with two rounds of 37mm AAA.

> "Two rounds detonated only seconds apart, the first just outside the aircraft and the second a direct hit on the fuselage by the LAU-74 Flare Launcher. It is believed there were no flares onboard, but there were flammable markers in the launcher area. The hit came in from 7 O'clock, causing immediate explosion, fire and smoke." Statement by SSgt Eugene Fields, Aerial Gunner, 24 April, 1970 (U).

Within thirty seconds, the aircraft began to fill with thick, black smoke. SSgt Fields made his way from the back to the front of the aircraft in complete darkness, unaware his intercom cord had become unplugged. When he reached the forward 20mm guns, the entire aircraft was filled with heavy smoke which caused him to blackout. When he came to, he didn't hear crew communications and assumed the rest of the crew had already bailed out. In the dark, he located a parachute on the front of the booth, clipped it on his harness, and bailed out of the right scanner's hatch. His exit was violent and as he bounced off the side of the aircraft, his helmet was ripped off his head. After a few seconds of free fall, SSgt Fields deployed his parachute. While descending, he saw his aircraft fly away in a straight shallow descent, trailing burning debris, until it crashed into a hill about 3.5 miles away. From a position overhead, Killer 01 observed the gunship impact the ground.

> "The AC-130 cleared us in on the gun and about halfway down the chute the 130 AC transmitted in a very cool voice "I'm hit, Babe." We asked him to repeat and he said "I'm hit." At this time, we were pulling off the pass and noticed the 130 was on fire. The aircraft appeared to be under control." Statement by Maj Richard Merkel and Maj Eddie Ryan, Killer 01 crew, April, 1970 (U).

Killer 01 remained overhead to search for survivors. Three to five minutes later, one emergency beeper was detected, but no voice contact was made. When Killer 02 arrived, voice contact with SSgt Fields was made and he described his condition. There was no radio contact or sign of any other survivors. The next day, SSgt Fields was picked up by aerial search and rescue (SAR) forces who continued to search for other crewmen throughout the day.

"On the morning of 22 April 1970, the 40th Aerospace Rescue & Recovery Service (ARRS, the HH-53 Jolly Greens Giants from Udorn RTAFB) responded to the downing of Warlord (Adlib 11) by sending one of its helicopters to shoot-down site. The Jolly was escorted by a sister ship and four A-1 series aircraft (call sign Sandy) to act as close air support. The known lone survivor's location was only a few klicks west of the Ho Chi Minh Trail, thus the pre-extraction work-up phase was lengthy, intense, and thorough. It was mid-morning before the Jolly was cleared into the extraction site. SSgt Fields was located in a depression within some foothills of bamboo and single-tier scrub brush and trees, a relatively safe E&E location. The smoldering impact point of Adlib 11 could be seen in the distance. Once the Jolly was escorted into Fields' general location, only a few minutes elapsed before he was positively located, and the jungle penetrator was lowered to him. He attached himself to it, and he was hoisted aboard. While the PJs treated his injuries, consisting primarily of second-degree burns and dehydration, the CSAR force thoroughly scoured the area for other survivors but to no avail. The Jolly rendered specific attention to the final

> descent path, surrounding area, and impact point of Adlib 11 but to no positive outcome. Finally, predicated upon the on-scene testimony of SSgt Fields and the absence of any emergency locator signals and/or parachutes--in addition to an increasing presence of enemy personnel and armament from the Trail--it was decided to vacate the area." **CMSgt (Ret.) Wayne Fisk,** Pararescueman, 40th ARRS.

When it was evident there were no other survivors, the crew of Adlib 11 was officially listed as Missing In Action (MIA). The reason the remaining ten crewmen aboard *Warlord* didn't bail out are unknown and remain a mystery today.[42] Though the loss of Adlib 11 was tragic and unexplained, AC-130 missions continued unabated until *Commando Hunt III* ended on 30 April 1970.

The month of May 1970 was plagued with spotty weather conditions and heavy AAA while searching for the few remaining trucks capable of operating on the trail. On 6 May, *Ghost Rider* received minor battle damage while searching but bad weather was a major factor. By the end of the month, the rainy season was in full force, making most of the Ho Chi Minh Trail impassable. This dwell in mission activity was the impetus for hasty AC-130 fleet upgrades

42 Initially, SSgt Fields used the wrong Search And Rescue Identification code, which caused confusion and hope for another survivor. The event was thoroughly investigated with no signs of other survivors, but unsubstantiated rumors persist that others may have survived and may have been captured by Pathet Lao.

before the start of *Commando Hunt V*, scheduled to begin on 10 October 1970.

Surprise Package (*Thor*) was touted by 7th AF as being the most potent AC-130 to date. Because of its digital fire-control, advanced avionics, 40mm guns, and higher altitude capability, Thor was sent into areas where trucks were plentiful, but well-protected by AAA. Regardless of the AAA threat, *Thor* turned in impressive truck kill numbers on nearly every mission. During its extended combat evaluation, it flew 112 missions, destroyed 604 trucks, and damaged 218. It was engaged by 105 rounds of 57mm, 32,271 rounds of 37mm, 8,541 rounds of 23mm, 25 unguided rockets, and 175 rounds of 14.5mm AAA. It was hit with shrapnel twice but was otherwise undamaged. *Thor* left Ubon for Wright-Patterson AFB, Ohio on 2 June for systems update and rework.

General Brown wanted more *Surprise Package* aircraft for the upcoming dry season, but *Thor* was a technology testbed, not a production aircraft. Thor's unique systems were hastily cobbled together for evaluation purposes and were not ready for fleet-wide integration. Furthermore, no technical manuals were available for Air Force maintainers, and many of the subsystems required specialized maintenance by civilian contractors. It was obvious ASD could not resolve all the issues in four months when all AC-130s were supposed to return to Ubon. Facing the dilemma, Secretary of the Air Force Dr. Robert Seamans directed the upgrade of all *Plain Jane* models with as many *Surprise Package* equipment modifications as possible; this project was known as *Pave Pronto*. About the same time, he recommended the pursuit of newer, more supportable C-130E aircraft for modification into gunships. As one might expect, there were differing opinions and discussion throughout the USAF chain of command all the way up to the Secretary of Defense, but the final direction was to modify all five *Plain Janes* (53-3129, 54-1623, 54-1627, 54-1628, 54-1630) into the *Pave Pronto* configuration, build additional *Pave Pronto* aircraft.

Basic AC-130 training at Lockbourne Air Force Base, Ohio was also an issue. The summer months were passing faster than the ability to adequately train crews, so some unusual accommodations had to be made. Since no AC-130s were available to train crewmen, they were given six sorties of instruction on an AC-119, then sent to Ubon. This less than satisfactory training method relied heavily on on-the-job training under combat conditions in Southeast Asia. Fortunately, all AC-130 pilots were senior officers who had significant experience in the C-130. Once at Ubon, new arrivals were trained via mentorship alongside seasoned crew members they were about to replace. As was done in the previous year, initial missions were flown in combat operations over permissive areas far removed from the Ho Chi Minh Trail. Nevertheless, after only a few training missions, crews were sent directly into the heavy AAA environment of *Steel Tiger East* to hunt and kill trucks. Once over the trail, crews discovered there was no shortage of targets or AAA fire. Intelligence sources estimated thousands of trucks were operating on the Ho Chi Minh Trail every night, and for *Spectre* crews, it seemed the PAVN had a never-ending supply of trucks and AAA.

> "As compared with 1969, the number of supply trucks belonging to Group 559 during the 1970-71 dry season increased from 3,000 to 6,500 trucks. The number of anti-aircraft artillery guns and mortars of all types was increased by over 200 tubes. The number of tractors of various types to build roads and to destroy mines and bombs increased by more than 1000. Communications radios were increased from 217 to more than 1000 sets. Troop strength was increased to 90,000 soldiers." PAVN History Records; English translation by Merle L. Pribbenow CIA Operations Officer (Ret.)

In September 1970, *Spectre* supported *Operation Tailwind*, a classified operation conducted by a US Army Special Forces MACV-SOG team of sixteen Americans and 110 Montagnard fighters.[43] Known as a "Hatchet Force," they were tasked to act as a diversionary force for a CIA mercenary force that was in trouble sixty miles to the north. The Hatchet Force was to be inserted deep into *Steel Tiger East*, directly into an area heavily controlled by the PAVN. AC-130 crews knew this area well, since it was adjacent to Route 65 of the Ho Chi Minh Trail, directly in one of the VR sectors they routinely worked. On 11 September,

43 The true story of Operation Tailwind, Major (Ret.) John Plaster USAR

the Hatchet Force boarded a number of USMC CH-53 Sea Stallions and headed for their objective. Escorted by twelve USMC AH-1 Cobra gunships flying at treetop level, Cobras suppressed active AAA sites. While en route, some helicopters were hit by small arms fire, wounding several Montagnards. Once the team was on the ground, it was apparent they were almost directly on top of a major PAVN logistics and control hub. The enemy was caught by surprise while the Hatchet Force quickly overwhelmed their positions. A search of the area revealed twenty earthen bunkers containing thousands of 122mm and 140mm rockets.[44] While a two-man Special Forces demolition team set charges inside each bunker, PAVN reinforcements were sent to the area. For the next seventy-two hours, the Hatchet Force was supported by an armada of air support to keep from being overrun. During the daytime, a fleet of Cobra gunships, A-1 Skyraiders, and F-4 Phantoms were overhead. At night, a tag-team of AC-130 and AC-119 gunships remained overhead at all times, engaging any enemy firing at, or maneuvering towards, the Hatchet Force. The force kept moving at all times to prevent being cut off by the enemy. At one point, the Hatchet Force attacked a headquarters element, killing fifty-four enemy troops. A search of the area revealed four vehicles and about nine tons of rice and a few munitions. They also discovered an underground bunker with maps on the wall and several hundred pounds of documents; the largest cache of enemy intelligence documents ever recovered in the war. By day three, the Hatchet Force was successfully

44 Ibid

withdrawn by Sea Stallions supported by US fighter aircraft. All sixteen Americans and thirty-three Montagnards were wounded and three were killed. The surprise attack on a major hub of the trail network was a monumental PAVN loss. During the operation, a major logistics encampment was destroyed and scores of PAVN soldiers were killed. After-action analysis of the captured documents revealed a great amount of detail relating to the operation of the Ho Chi Minh Trail network, which was exploited extensively by 7th Air Force Intelligence analysts.

By October 1970, *Pave Pronto* AC-130s were arriving at Ubon in a vastly different configuration than the *Plain Jane* model. Equipment additions included the AN/ASD-5 Black Crow sensor, inertial navigation system (INS), and LORAN, which enabled precise and accurate location of navigation points and initial points of road networks. The INS eliminated the need for flare drops and visual NOD/IR searches to the fix initial point (IP) at the beginning of a search. The aircraft was equipped with a commercial video recorder that eliminated the need for a secondary NOD and aerial photographer. Critical flight controls, hydraulics, and crew stations were protected by ceramic armor plate, a pair of 40mm M2A1 modified guns replaced aft 20mm and 7.62mm guns, and a new analog fire-control computer was installed. One thousand rounds of experimental 40mm high explosive incendiary (HEI) "Misch Metal" rounds were sent for a combat evaluation. These rounds were lined with a pyrophoric metal sleeve similar to cigarette lighter flint that produced thousands of burning fragments upon detonation.

The NOD was replaced with a low-light-level television (LLLTV) sensor mounted in the former crew entrance door. A laser target designator (LTD) was added to the LLLTV gimbal to designate targets for fighters loaded with laser-guided bombs. The LLLTV sensor operator, IR sensor operator, and Electronic Warfare Officer (EWO) were located in the sensor booth. The EWO operated the Black Crow sensor to track electrical emissions from the gasoline-powered engines. Though the Black Crow was not accurate enough to precisely locate a vehicle, it narrowed down the search area and provided cueing for the visual sensors. The EWO also operated the aircraft electronic countermeasures (ECM) systems on the aircraft to include the newly installed ECM pods on each wing.

> "Arrival of the Black Crow sensor drastically changed our initial target acquisition. No need to corkscrew down the trails (LOCs) or rely on the iffy information from Igloo White ground sensors relayed thru Batcat by Task Force Alpha. Approaching your assigned strike box, Black Crow gave you an 8-19 nautical mile bearing to non-diesel ignition vehicles. Ability to accurately fire off Black Crow was not near as accurate as other sensors." **MGen (Ret.) Bob Patterson,** pilot.

Commando Hunt V began on 10 October 1970 under heavy clouds and persistent rain. Since most of the trails in *Steel Tiger* remained impassable, truck traffic was minimal. Newly arrived *Pave Pronto* crews flew familiarization missions into lightly defended areas of Laos and Cambodia to gain practical combat experience. These missions were beneficial training for newly arrived crewmen who would soon find themselves in intense combat operations in *Steel Tiger East*.

> "In November, 1970, I attended AC-130 ground school at Lockbourne. I flew 5 missions in C-119K's and only two were 20MM live-fire ... never saw an AC-130 until I got to Ubon mid Nov 1970. I flew AC-130A training missions in Cambodia to become mission qualified." **Col (Ret.) Jim Kyle,** Pilot.

Sgt Don Dawson loading 40mm HE ammunition on a Pave Pronto aircraft. (USAF Photo)

By the middle of November 1970, the weather had improved enough to schedule two or three missions per night into *Steel Tiger East*, but not without significant aircraft maintenance difficulties. Newly installed *Pave Pronto* avionics, especially the analog fire-control system, were finicky and failure-prone. There was always at least one aircraft grounded for long-term repairs, which, of course, fell to further disrepair by cannibalization of parts for other aircraft. Maintenance delays were common not only because of unique, complicated avionics but because of the aged aircraft themselves. Everything about the AC-130A base aircraft was antiquated or obsolete right down to the three-bladed Aero Product propellers, which were not used by any modern model C-130 in the USAF.

A mission that typified crew frustration occurred on 26 November 1970, when Capt Hight's crew (call sign Bixby) was delayed for more than two hours for maintenance before finally taking off. Once over target, they found numerous trucks but had difficulty scoring hits. After spending their entire mission troubleshooting the fire-control system, they returned to base low on fuel and ammunition with zero truck kills and only a few damaged. Unfortunately, these failures were becoming all too common with the *Pave Pronto* analog fire-control computer.

> "From a crew point of view, it was a mission of pure frustration. When targets were found and the pilot attempted to roll into firing geometry, several things might happen,

> and one or more usually did. The aircraft might not fly proper geometry if the command guidance from the computer was followed. Once in orbit, if the "pippers" were followed, the aircraft would "screw out" or "screw in". If the pilot fired with pippers superimposed, the shot would be wild with no apparent correlation to anything. Add a 30-knot wind to this mixture and the recipe for frustration was complete." 16 SOS History Records, Oct-Dec, 1970.

From the beginning of the *Pave Pronto* program, there were fire-control system problems. Crews attempted to work around the issues but spent far too much time wrestling with the system while exposed to AAA fire instead of engaging targets. An urgent request for assistance was sent to the Gunship Program Office and a troubleshooting team consisting of Lt Col Terry, Lt Col Krause, Wg Cdr Thomas Pinkerton, Maj Clifford, and other systems experts traveled to Ubon. The team brought not only systems expertise, but technical data, spare parts, and test data. One of the most important items of data were the computer run-outs which gave more accurate gun settings. Aircraft fire-control accuracy was analyzed by experienced crews during daylight missions in permissive areas. Within a short time, most of the gross error issues were resolved and procedures were developed and documented for crew coordination. Multiple problems with fire control, gun sight, autopilot, and gyros were encountered and field repairs were made.

> "Lt Col Ron Terry and team came to Ubon with a new computer and offered technical assistance. I took off with Col. Terry, his new computer installed and drove directly to a destroyed bridge in the Mekong River that we used daily to boresight and align the guns. Ron Terry had an unbelievable awareness. Standing behind my pilot seat, it was just like he could see through the gunsight."
> **Maj Gen (Ret.) Bob Patterson,** pilot

Through a system of troubleshooting, the cause of poor performance was found to be poor workmanship on the fire-control system at the CONUS depot.[45] The Gunship Program Office repaired the system but recognized a lack of adequate training, technical manuals, and data was still a problem. To remedy the issue, special pilot, nav, and FCO training sessions were held to discuss techniques of how to fly precise geometry to attain the proper gunsight picture and tactics to compensate for wind effect. Gradually, aircraft became available and crews began to learn, but many problems were encountered. The fire-control system seemed to go amiss, bore sighting temporarily became a lost art and still, the weather did not break to enable in-flight training. The training challenges were steadily overcome and crews became proficient. By early December, the trails were dry again and thousands of trucks operated on the trail every night. *Spectre* crews continued to patrol the trails, firing on every truck they could find.

45 E-mail from Col (Ret.) Jim Kyle.

After several years of relentless AC-130 attacks, most *Group 559* drivers felt there was no escape from the AC-130, so they usually stopped in the middle of the trail and ran into the jungle to avoid being killed. The situation worked well for *Spectre* crews since it was easier to engage stationary targets. As a result, trucks damaged or killed by the AC-130 skyrocketed, causing a major disruption of *Group 559* supply efforts. When word of the situation reached *Group 559* commander General Dong Si Nguyen, he was shocked at the number of trucks and supplies he was losing to the AC-130. When he learned drivers abandoned their trucks whenever an AC-130 attacked, he became angry and accused his drivers of cowardice. Drivers were adamant there was no defense against the AC-130, claiming the aircraft could see at night, fire on them for hours at a time, and that staying in the cab of their truck was certain death. General Nguyen declared there was no such capability, and the drivers were exaggerating. To prove his point, he announced he would accompany the lead driver of a fifty truck convoy scheduled for that evening.

> "This is unacceptable. If drivers are going to die, they should die on the steering-wheels of their trucks. To have our officers and men scatter like a flock of geese just because of one Old Lady [PAVN nickname for the AC-130] is an insult to our intellect, to our intelligence. Tonight, you are to organize a 50-truck convoy from the 102nd Battalion

> to drive down the exact same stretch of road where that incident happened last night. I am going to ride beside the lead driver and personally command this convoy." Memories of the Ho Chi Minh Trail, (Ky Uc Truong Son), General Dong Si Nguyen. English translation by Merle L. Pribbenow CIA Operations Officer (Ret.)

That night, the weather was clear and the moon was bright as a fifty truck convoy led by General Dong Si Nguyen set out in blacked-out trucks, camouflaged with fresh-cut vegetation. As the convoy proceeded down a winding mountainous jungle road, all seemed to be going well. About midway through their journey, they heard the familiar sound of AC-130 propellers. Capt "Hop" Riopel's crew (call sign Spur), escorted by three F-4s (call sign Dudley Flight), sighted the convoy and established a firing orbit 8,500 ft. overhead. Within minutes, Spur knocked out the lead truck, then the trailing truck, trapping all vehicles in between. There was no escape for any of the trapped vehicles. For more than an hour, Spur engaged the convoy, with 276 rounds of 40mm, and 3,000 rounds of 20mm, ultimately destroying 31 trucks and damaging 2. Dudley 01 also destroyed two trucks. During the attack, General Nguyen and all drivers remained in their trucks to their own peril. During this time, only one 37mm gun opened fire on Spur, firing a total of 21 inaccurate rounds. When Spur departed, they felt it was just another good night of truck hunting, but from a strategic perspective, it was much more. The effect felt by General Nguyen was a significant watershed moment for *Group 559*. In his Headquarters Command Post the next day, a wounded General Dong Si Nguyen stood up and declared:

> "The fault lies with me, not with our drivers. Our men are brave and resolute, but bravery and resolve alone will not defeat the enemy. I underestimated the Americans. They are able to put a man on the moon, which means they can do a lot of other things with their technology. The truth is that the AC-130 is fitted with very sophisticated surveillance equipment. It can detect our trucks even before the trucks arrive in its area of operations. Our problem now is that we must immediately institute new measures and come up with new ways to carry out our duties. We must re-deploy our anti-aircraft forces...." Memories of the Ho Chi Minh Trail, (Ky Uc Truong Son), General Dong Si Nguyen. English translation by Merle L. Pribbenow CIA Operations Officer (Ret.)

General Nguyen's drive to keep supplies flowing into South Vietnam was greatly hindered by *Spectre* and he was frustrated the PAVN air defense network was unable to shoot down an AC-130. In response, he encouraged his staff to develop tactics and procedures to defeat what had turned into *Group 559*'s archnemesis.

By 1 January 1971, nine Pave Pronto and the *Surprise Package* aircraft (*Thor*) were on station at Ubon.[46] Since Capt Riopel's mission earlier in the month deeply affected General Nguyen, he ordered his AAA gunners to shoot down AC-130s at all costs. A bounty was placed on *Spectre* crewmen and AAA crews were promised rewards of cigarettes, food, and luxuries to anyone who shot down an AC-130. The word spread quickly within PAVN ranks. Soon, US intelligence intercepted PAVN radio transmissions and the 7th AF HQ intelligence posted an advisory indicating increased efforts to shoot down an AC-130 on or near intersections on Route 9, west of Tchepone.

0 110830Z DEC 70
FM 7AF TAN SON NHUT AFLD RVN
TO RUMOUBA/8 TAC FTR WG
UBON RTAFT THAI
SUBJECT: AAA WARNING ADVISORY (U)

USUALLY RELIABLE SOURCES REVEAL THAT THE ENEMY MAY TAKE INCREASED EFFORTS TO SHOOT DOWN AC-130 GUNSHIPS ALONG ROUTE 9 IN SOUTHERN LAOS. AIR CREWS SHOULD BE CAUTIONED OF THE THREAT IN THIS AREA PARTICULARLY AROUND ROAD JUNCTIONS ALONG THE ROUTE.

GP-3 BT (16 SOS History Records, Dec, 1970)

46 Tail Numbers (in order of arrival) were: 53-3129, 56-0490, 54-1627, 54-1628, 55-0029, 54-1623, 54-1630, 55-0046, 55-509, 55-0040. From Jan-Mar, 1971, four more AC-130 Pave Pronto aircraft arrived at Ubon: 56-0469, 55-0044, 55-0043 and 55-0014.

North Vietnamese photo believed to be General Dong Si Nguyen while he visited with 37mm AAA gunners somewhere in Laos. (Photo credit unknown)

Commando Hunt V was about half complete when the 16 SOS was permitted to use their original call sign Spectre. The effort to stop the traffic flow on the Ho Chi Minh Trail continued, limited only by weather conditions and aircraft availability. Hundreds of fighter and bomber missions were flown into Laos and Cambodia. During the day, fighters and bombers continued to attack trail intersections with delayed fuze bombs to create deep craters and make them impassable. In most cases, the effect was temporary since trail workers detoured traffic onto parallel trails while repairing the bombed-out areas. In mountainous areas where detours onto other trails were not possible, trail workers built camouflaged bypasses around bomb craters but did not repair them. This tactic gave the appearance the road was impassible; but, in fact, it was a serviceable trail. Most of the time, trail workers worked in daylight when the trail was essentially void of traffic, but there were times when repairs were done at night. Bulldozers and heavy equipment were rare commodities that were targeted the same way as trucks when discovered, so much of the work was done by hand with primitive tools.

Spectre crews continued missions unabated. Of high concern were frequent avionics failures that continued to hinder mission accomplishment and increase vulnerability to enemy fire. Most critical to the effectiveness and survival of the AC-130 was the amount of time it took to find, target, engage, and assess damage. Since the early days of AC-130 combat operations, crews determined the first rule of survival was "limit exposure" and the second rule was "always expect to be fired upon when firing." Crews were concerned since all too often they were launched with only one operational sensor, needlessly exposing them to hostile fire for extended periods with limited ability to fight back. If all three sensors were operational, the *Black Crow* was used to search a wide area, detect ignition signatures and provide general guidance to the pilot. Once the BC detected electrical emissions, the IR sensor would be slaved to the BC to establish a firing orbit, then visually acquire the target(s). When all three sensors worked as designed, less time was needed to engage targets and overall kill rates were higher. Launching an AC-130 with only the BC was unwise since the system was too coarse for precise targeting. The LLLTV required some level of illumination to operate properly and had a limited field of view because of its position in the former crew entrance door opening. In the end, the IR sensor was singled out as the most important visual sensor since it required no ambient light to operate and had a wider field of view than the LLLTV. In January of 1971, Lt Col George Fox, 16 SOS assistant operations officer, sent a letter to the 7th Air Force Director of Operations concerning the vulnerability of the AC-130 when operating with degraded sensor capability. Considering the factors of risk versus gain, an agreement was reached to not launch an AC-130 without an operational IR sensor.

The first big mission of 1971 was flown on 14 January by Maj Edmonson, on *Proud Warrior* (call sign Spectre 04). While en route to a routine mission into area 11 of *Steel Tiger*, they first rolled into an orbit to tweak their guns in area 10, a known permissive area of Laos. Sensor operators were surprised when they stumbled on what appeared to be an unprotected truck park. The crew immediately began to tweak their guns, making corrections on the fly, until they were getting direct hits and starting fires. The mass engagement made them late for tasking into their assigned area. When questioned by *Moonbeam* why they were behind schedule, Spectre 04 advised *Moonbeam* of their find and requested to remain over the target to continue their attack. For the next three hours, Spectre 04 and their escorts Outfox Flight fired on targets below. This was a prime example of mission flexibility insight and the "bird in the hand" methodology by *Spectre*. During this mission, the 40mm Misch Metal round was found to be "almost magical" in starting fires. Other crews found the new round equally impressive, so an urgent 7,000 round order was placed for Misch Metal. The effect of burning trucks and wholesale death and destruction on the trail intimidated drivers to the point some avoided transporting weapons ammunition and fuel.

> "Beginning with the 1970-1971 dry season, the enemy changed tactics, switching from trying to destroy bridges and roads to making attacks aimed at our personnel. The attacks were aimed primarily against our trucks, our anti-aircraft artillery positions, our warehouses, etc. During this period, the Ho Chi Minh Trail's most dangerous opponent was the AC-130. Because we were overoptimistic and even inebriated with our victories, we were slow to recognize and study the enemy's new tactics and equipment. When the number of trucks destroyed and drivers killed or wounded began to grow day by day, a number of cadre and drivers became concerned and frightened. According to reports received from a few of our way stations [Binh Tram], some drivers tried to avoid transporting weapons, ammunition, and fuel. This was something that had never before occurred in the history of the Ho Chi Minh transportation corps." The Road Through The Annamite Mountains: A

Memoir, [Duong Xuyen Truong Son: Hoi Ky]
English translation by Mr. Merle Pribbenow,
CIA Operations Officer (Ret.).

The human factors element for both *Spectre* crews and *Group 559* drivers was remarkably similar. While gunship crews spent the bulk of their off-duty time preparing to fly their next mission, not all missions were eventful and some were nothing more than hours of boredom without finding any targets. *Group 559* truck drivers found themselves in a similar quandary. Most of their time was spent in garrison loading supplies and waiting for darkness to complete their nightly round trip. Transits were routine and dull until an AC-130 arrived overhead, which caused considerable fear and panic. Sometimes drivers would abandon their trucks on steep terrain only to have them roll downhill and hit other trucks or tumble off the side of a cliff. Once the AC-130 departed, drivers would regroup and walk back to their Binh Tram while other trail workers arrived to recover whatever supplies they could while trail workers towed damaged vehicles back to their Binh Tram for repair. While PAVN commanders were disappointed with their inability to shoot down an AC-130, AAA gunners were demoralized and universally frightened of the AC-130 and their fighter escorts.

The dilemma was personally distressing to General Nguyen, who was nearly killed by an AC-130, lost thousands of trucks, tons of supplies, and many personnel to *Spectre*. Horrific stories circulated from Binh Tram to bihn tram about drivers who met their fate trying to evade the gunship. The level of folklore was often exaggerated when told and retold by PAVN soldiers, resulting in an overestimation of AC-130 capabilities. Discounting runaway folklore, General Nguyen focused on an effort to collect intelligence on exactly how the aircraft functioned as a way to devise defenses against it. Finding credible information was next to impossible, so he directed his forces to concentrate efforts to shoot down an AC-130 to examine the wreckage. To accomplish this, they devised a plan to trick an AC-130 crew into a AAA ambush.

> "The issue was now clear. Group 559 Headquarters concluded that the AC-130 aircraft was an extremely dangerous foe with respect to our nighttime truck transportation operation. The factor that made the AC-130 the most dangerous was its superior ability to stay in the air for an extended period of time, flying in small circles to create a "mobile fortress." These aircraft could take turns suppressing and destroying our truck convoys. This meant that the entire supply corridor had to find an overall solution that was creative, flexible, secret, and unexpected." Memories of the Ho Chi Minh Trail, (Ky Uc Truong Son), General Dong Si Nguyen. English translation by Merle L. Pribbenow CIA Operations Officer (Ret.)

While North Vietnamese intelligence made a pointed effort to research capabilities of the AC-130, many potential solutions to mitigate risk were considered. Drivers experimented with tricks of their own, but the capabilities of the gunship seemed mysterious and almost magical. Commanders left their headquarters to visit AAA sites, road watch teams, and truck drivers. Understandably, driver observations varied widely. Some drivers were not fearful of the AC-130, but they were in the minority. The majority of rank and file drivers considered the AC-130 a beast that destroyed everything it could see and pounded them relentlessly for hours at a time. They learned by witnessing fellow drivers being killed, burned, or blown to pieces as their trucks were hit by *Spectre* gunfire. Many felt the only way to escape certain injury or death was to avoid detection altogether. Some felt engine heat was detectable by AC-130 sensors, so they strapped heavy bamboo fiber mats or wet blankets on their hoods to mask infrared signature. Many drivers realized the gunship could find any truck with a running engine and destroy it in complete darkness. Some considered turning their engines off whenever an AC-130 was nearby to avoid detection, but sections of the trail would be shut down for several hours whenever an AC-130 was nearby. Suspecting electrical emissions from their trucks were being used for targeting, enterprising drivers wrapped aluminum foil around their ignition coils and spark plug wires.

> "He (Colonel Xi) met with various section leaders, who told him of the dam-

> age the AC-130 were doing. When he met with numerous drivers and asked for their thoughts. One stated that he hated the lumbering gunships and their massive guns, which he called the "thug." Another driver observed: When we stayed quiet, it [the gunship] merely circled overhead. But as soon as we started the engine, it began firing. If we turned off the engine, it would fire a few more rounds and leave. But if we restarted the motor, it would come back at once. The dangerous thing is that this plane needs neither flares nor guidance by a recon plane" *War in Peace, Vol. number 5, 1984, Road to Victory: The Ho Chi Minh Trail.*

Knowing the AC-130 had extreme difficulty finding or firing on targets in heavily camouflaged Binh Trams, caves, or under triple canopy jungle, Gen Nguyen and his headquarters staff saw complete stealth as their only real defense against the AC-130. To accomplish this, he envisioned the construction of a secret trail to be cut through a heavy canopied jungle for its entire length, thus masking movement and detection by the AC-130. From that point on, a tremendous amount of manpower and equipment was devoted to constructing the secret trail, which would take shape as the K-Trail during the following year. While the K-Trail was under construction, *Group 559* took other steps to avoid exposure to AC-130 attacks. The 591st Anti-Aircraft

Artillery Regiment established numerous 57mm AAA sites along the trail and increased efforts to trick AC-130 crews into AAA traps. To further their effort, trail workers ratcheted up efforts to build large fake convoys ringed by dozens of camouflaged AAA guns. Sometimes they would place burning embers under the engines of shot-up trucks and even run small gasoline engines to lure AC-130s into AAA traps.

Another defense used was known as "lead-truck decoy." This tactic equipped an especially courageous lead driver with a lightly armored, highly maneuverable vehicle. As soon as the AC-130 would arrive overhead, the convoy would stop while the lead driver would speed ahead hoping the AC-130 crew would chase it into an AAA trap. If the gunship crew took the bait and pursued the lead vehicle, the rest of the convoy would turn around or vanish undercover. In practice, *Spectre* crews caught onto this trick quickly.

> "With regard to our transportation operations, we would have to use a combination of three different methods. Large truck convoys would travel by day down the secret road. Small truck convoys would travel at night down the open, overt road. We would also select brave and intelligent cadre and drivers to organize a number of small squads of trucks to drive along "decoy" roads. Their job would be to serve as decoys to distract and lure in the AC-130s so that our anti-air-

> craft guns could engage and destroy them."
> Memories of the Ho Chi Minh Trail, (Ky Uc Truong Son), General Dong Si Nguyen. English translation by Merle L. Pribbenow CIA Operations Officer (Ret.)

After General Nguyen's fateful convoy attack, *Spectre* crews noticed a drop-off in truck traffic until a short period just before sunrise. Once *Spectre* departed, drivers knew they had only a brief period of time, maybe only thirty minutes, to speed from one Binh Tram to the next before fighters and bombers arrived overhead. Drivers were initially fearful they would be attacked by tactical aircraft, but found they were rarely attacked by fighters or bombers. In response to *Group 559*'s new PAVN tactic, daylight AC-130 missions were flown at 15,000 ft. using the Black Crow ignition detection system to locate convoys. Electronic Warfare Officers found the effort only marginally effective. The *Black Crow* system did not work well at high altitudes, accuracy suffered greatly, and the aircraft was more vulnerable to optical 57mm AAA. Since the concept proved ineffective, daylight missions were discontinued after about two weeks.

Back on a strict night-time schedule, *Spectre* continued hunting trucks in *Steel Tiger* East and *Barrel Roll* West. Most missions met some level of accurate 37mm AAA, even at *Pave Pronto's* higher operational altitude, and it was clear PAVN gunners were getting more proficient targeting *Spectre*. On 21 January 1971, *Prometheus* was hit by 37mm in the leading edge of the right wing. It was repaired fairly quickly, but two nights later it was damaged again by 37mm AAA. Two nights later, *Ghost Rider* was working around a scattered low undercast when a burst of 37mm AAA came through the clouds hitting its number one engine. Fortunately, the crew successfully recovered at Ubon. A little more than a month later, *Prometheus* was hit again by 37mm and fragments shattered the copilot's side window. By then, crews were referring to *Prometheus* as an "AAA magnet." About a month after that, *The First Lady* was hit by AAA for the fourth time.

In early 1971, while combat operations continued on the Ho Chi Minh Trail, all AC-130 initial training responsibilities were transferred from Lockbourne AFB, OH, to the 415th Special Operations Training Squadron at Hurlburt Field, FL. During the same period, Secretary of the Air Force Robert Seamans, Jr. directed funding for *Project Pave Spectre*; an effort to build eleven more AC-130s advocated by former AC-130 crewmen Lt Col "Gunship Charlie" Spicka and Lt Col Charles Gentzel, both assigned to the Pentagon. The AC-130 had made its mark in combat and was considered a high priority in the effort to sever the Ho Chi Minh Trail. With this level of support, ASD was authorized wide latitude to develop the "ultimate AC-130" by modifying nearly new 1969-year production C-130E aircraft. The C-130E had a longer range, modern four-bladed propellers, more power, and modern avionics and navigation systems. The rush was on to get as many AC-130E gunships as possible built and delivered before *Commando Hunt VII*, scheduled to begin on 1 November 1971. At about the same time, the

Gunship Program Office and 4950 Test Wing (Tech) brainstormed ideas to increase both lethality and stand-off range of the AC-130. They studied the potential for mounting a 75mm and 105mm gun, ultimately settling on the US Army's M102, 105mm Howitzer. This effort, named *Project Pave Aegis* was an in-house engineering and system development effort managed by the Gunship Program Office at Wright-Patterson AFB, Ohio, and tested at Eglin Air Force Base, Florida, and Dahlgren Navy Base, Virginia. ASD engineers planned to have an operational system ready for combat evaluations by the start of the 1971 dry season.

Back at Ubon, on 3 February, the crew of *Grim Reaper* (call sign Spectre 12) used its Laser Target Designator (LTD) to designate a 37mm gun site for its F-4 escort, who destroyed the gun with a laser-guided bomb. This was the first time an AC-130 used its LTD in combat. Beginning on 8 February, numerous AC-130 missions were flown in support of Lam Son 719; the largest ARVN ground operation of the war. The objective of this operation was to sever the Ho Chi Minh Trail near the strategic chokepoint of route 9 near Tchepone, Laos, and cripple North Vietnam's vital Ho Chi Minh Trail supply route. This strategic chokepoint intersection of roads and waterways was well known to *Spectre* crews since it was the same location where *Prometheus*, *Grim Reaper*, and *Ghost Rider* were hit by 37mm AAA in January. The PAVN had good reason to defend Tchepone since it was considered a strategic nexus near Base Area 604, General Dong Si Nguyen's headquarters, and the main PAVN logistical transshipment area in Laos. Route 9 was a direct connection into South Vietnam, just west of Khe Sanh, Vietnam. US Intelligence estimated twenty-two thousand PAVN soldiers, support personnel, and Pathet Lao operated in Base Area 604. Therefore, disrupting PAVN activity there would shut down the entire Ho Chi Minh Trail. Though the plan was considered risky, ARVN commander Lt Gen Hoang Xuan Lam ordered the operation to commence. *Spectre's* role in Lam Son 719 was to provide ARVN forces with fire support and interdict PAVN troop reinforcements, supplies, and equipment.

When ARVN airborne and armored forces began to probe into Laos, initial PAVN resistance was light as ARVN established artillery firebases and forward operating posts. The operation quickly soured when the ARVN command network and battle plan bogged down. During this time, PAVN commanders infiltrated thousands of troops into the area to reinforce Base Area 604. By late February, more than twenty thousand ARVN armored and infantry soldiers had crossed the Laotian border, supported by US airpower and elements of the US Army and Marine Corps. By that time, ARVN troops battled more than thirty-five thousand PAVN troops in and around Base Area 604, resulting in heavy losses on both sides.

While PAVN and ARVN troops engaged each other in fierce combat, AC-130 crews were overwhelmed with mass PAVN reinforcements and supplies on the trail. Traffic was so heavy crews couldn't spend a lot of time attacking any one vehicle to ensure total destruction by fire. Instead, the sheer number of targets necessitated literal interpretation of 7th Air Force approved truck-kill criteria as one direct 40mm hit equals a kill. Though crews were aware the trucks were not likely destroyed unless they burned, the high number of targets and time needed to engage them was a major factor. *Spectre* crewmen estimated at least one out of four trucks were actually destroyed by fire, while the rest were considered damaged. Nevertheless, under current BDA criteria, there were reported as destroyed.

The PAVN took advantage of the chaos of *Lam Son 719* to establish additional 37mm and 57mm AAA sites throughout *Steel Tiger East*. As if the AAA wasn't enough, the PAVN 67th Missile Battalion established an SA-2 Surface-to-Air Missile (SAM) in an unusually flat and open spot near Tchepone, west of Base Area 604. This location was chosen as a suitable launch site since there was enough cleared area to situate and support the large SA-2 system. Remarkably, the entire system was transported and set up without being discovered by US reconnaissance or intelligence entities. Commonly referred as the "flying telephone pole," the SA-2 was equipped with a 440 lb. blast and fragmentation warhead that was capable of knocking out any aircraft in the US arsenal. A typical SA-2 site was comprised of three to six missiles, a launch control van, multiple support vans/vehicles, three-phase electrical generators, diesel fuel, liquid rocket fuel, a *Spoon Rest* (P-12) early warning radar, and a *Fan Song* fire control and tracking radar. Launchers holding the 5,100 lb., 35-foot long missiles were spaced several hundred yards apart, usually in a star pattern. Soviet advisors were typically assigned to missile batteries to train and supervise PAVN missile crews. PAVN training levels varied, but, even with experienced crews and Soviet advisors, hit probability was only about 30-50 percent. To increase effectiveness, either two or three missiles would be fired, spaced six seconds apart. A successful launch and guide-to-hit were only achieved by a high level of coordination, teamwork, and proper interpretation of radar signals. The physical capabilities of the Mach 3.5 missile were achieved by a solid-fuel booster and a larger liquid fueled missile motor.

The SA-2 employs a *Spoon Rest* wide-area search radar (aka P-12) to locate targets, and two "pencil beam" *Fan Song* fire control and guidance radars to guide missiles to target. *Fan Song* radar operator's television display was essentially an electronic version of looking through a soda straw. Because both radar operators had to superimpose their narrow tracks to achieve lock-on, it normally took several minutes before a lock-on could be made. Once launched, radar operators needed to keep tracking until the missile reached the target. Though an AC-130 crew could not detect *Spoon Rest* radar signals, *Fan Song* radar signals were detectable by the gunships Radar Homing And Warning (RHAW) receiver. Since it normally took several minutes for SA-2 launch officers to acquire, track and engage an aircraft, RHAW indications gave aircrews at least some time to apply countermeasures and maneuver to either avoid or break lock-on. If an SA-2 was launched, the visual signature was unmistakable as both solid rocket booster and liquid-fueled sustainer motors lit the jungle as if it was daylight. Since the

SA-2 was a serious threat, the vicinity of all known sites was off-limits to AC-130 crews.

On the night of 2 March 1971, *Midnight Express* (call sign Spectre 03), commanded by Maj Ed Holley, left Ubon for an armed reconnaissance mission in *Steel Tiger East* VR 4. Because of *Lam Son 719* activity near Base Area 604, Spectre 03 was not allowed to fly east of the town of Tchepone. Rolling into an orbit a few miles west of Tchepone at 9,500 ft., IR sensor operator Capt Lee Schuiten and LLLTV sensor operator Maj Ed Coogan searched for targets. Coogan was on his first AC-130 combat mission and being instructed by Maj Henry Zeybel, who stood behind, mentoring him. Meanwhile, the right scanner and IO began calling heavy, but inaccurate 37mm AAA. With each orbit, AAA would get closer causing the crew to periodically break out of orbit to avoid being hit. Within a short period, Maj Coogan had located a sizable convoy. Maj Dick Kauffman, the crew's navigator validated the targets, ensuring no friendly forces were nearby while gunners stood by their guns, ready for an aggressive attack. Henceforth, Maj Holley engaged the lead truck first, setting it on fire, then attacked the trail truck, trapping all other trucks in between.

> "I told Ed Holley he was cleared to fire since we were in a clear firing area. Henry was Ed Coogan's TV instructor. I put the TV into the Fire-control computer and set the gun fire window to 1 mil. Once we starting to fire Ed Coogan gave me the missed distance for each

> round, like 4 high, from the truck he was tracking. After about 5 rounds I had computed the wind between the aircraft and the target and had inserted it into the computer. The next several rounds hit the truck. After destroying that truck, we moved on to the next truck." **Lt Col (Ret.) Richard Kauffman**

For nearly two hours, Spectre 03 picked the convoy apart, destroying or damaging fifteen trucks. As the vehicles burned and exploded below, EWO Maj Jim Ballsmith advised the crew he was tracking two small groups of trucks with the Black Crow system. Maj Holley rolled out on BC guidance until they were about five miles south of Tchepone, just south of Delta 12. When Spectre 03 rolled in on the first group of vehicles, sensor operators located stationary vehicles with engines running, lined up in two closely spaced groups. As soon as Maj Holley established a firing orbit, multiple 37mm guns opened fire in what seemed to be inaccurate barrage fire. Maj Holley remained in orbit and fired one 40mm round before the aircraft's Radar Homing And Warning (RHAW) system sensed enemy radar activity. Immediately, Maj Ballsmith announced, "I have a SAM activity light." Without hesitation, Ballsmith directed the pilot to turn to heading 270 to put the radar behind them. Maj Holley rammed the throttles forward to increase airspeed and a few seconds later Maj Ballsmith advised the crew he had a "SAM ready to launch light." Though there may have been some level of doubt if they were being tricked by PAVN deception tactics, there was no doubt whatsoever when Maj Ballsmith called, "LAUNCH." SSgt Bob Savage, the IO, positioned on the open ramp saw a bright flash about ten miles behind the aircraft. The entire area below the aircraft was illuminated as a trail of fire ascended towards Spectre 03. SSgt Savage saw the missile's second stage ignite and directed the pilot to turn right, then left, then break left, but the missile tracked aircraft movement and kept closing in. SSgt Savage advised the crew the missile had leveled off and looked as if it were coming right in the ramp, then called, "DIVE, DIVE, DIVE!" Maj Holley said, "hang on, we are going down," and immediately rolled the aircraft into a 135-degree bank, then pointed the nose towards the ground until the aircraft was descending at a 4,000 feet-per-minute rate.

> "The maneuver was like turning over a dump truck and still expecting to steer it. Our turkey was not designed for such a feat. Yet Holley did it. Maintaining a positive one-gravity load on a creaky, antiquated airframe, he plunged downward in open defiance of several laws of aerodynamics."
> **Lt Col (Ret.) Henry Zeybel**

While *Midnight Express* descended, the missile kept coming. Maj Kauffman watched the unwinding altimeter, calmly calling out altitudes in 500-foot increments to prevent panic and subsequent collision with the ground. As the aircraft reached 3,000 ft., Maj Holley began to pull out of the dive, eventually leveling off at about 2,000 ft. Fortunately, the SA-2 could not maintain track of the rapidly descending AC-130 and it flew overhead and detonated well past the aircraft. Maj Holley elected to exit the threat area to better determine the status of the aircraft.

> "I was just totally focused on getting the plane down below the radar and getting the crew back safely. The hardest part was knowing when to come out of the dive. I had to rely on my instincts and the instincts of the navigator. The entire crew had to anticipate what was needed." **Maj Ed Holley**

The entire SA-2 engagement and maneuver had taken about 45 seconds, but it felt like an eternity to the crew. Now a good distance from the site, Maj Holley climbed back to 9,500 ft. and directed the crew to check in with their status. Remarkably, nobody was injured, but it took a while to get the aircraft's electrical generators stabilized. While they orbited, the crew discussed what had just occurred and agreed it was definitely an SA-2 engagement. SSgt Savage noted the missile reacted exactly how he was trained and Maj Ballsmith was the 16 SOS chief EWO who had previous experience with SA-2s. Once all crewmen had checked in and the aircraft was determined undamaged, Maj Holley queried his crew for a vector on follow-on activity. Considering the convoy was larger than normal and figuring there were no missiles remaining, the entire crew elected to return to the convoy to finish what they had started.

Spectre 03's notion was the missile was just a fluke to chase them away and they thought it was unlikely another missile was there. Ten minutes later, Spectre 03 returned, established a firing orbit, and began to engage the trucks. After hitting one truck, Maj Ballsmith advised the crew he had another SAM activity light, then directed a turn to a 300 heading. Within seconds, the situation unfolded almost identically as the first engagement and two SA-2s left their launchers. Maj Holley performed the same maneuver a second time. Fortunately, both missiles missed by a wide margin, detonating a safe distance away. As if one SA-2 engagement wasn't bad enough, two engagements convinced them going back was a bad idea. Suspecting the aircraft structure may have been damaged by aggressive maneuvers, Maj Holley elected to return to base. A short time later, Spectre 03 landed at Ubon without incident. Once safely on the ground, Maj Coogan turned to Maj Zeybel and asked him, "are all missions like that?" Maj Zeybel replied with dark humor "no, some missions are worse!"

Spectre 03 crewmen were certain the launch site was in a relatively open area near the Banghiang River, southwest of Tchepone. This area was placed off-limits to AC-130s while reconnaissance aircraft searched for the site. Suspecting more SA-2 sites would spring up soon, 16 SOS commander, Lt Col Ken Harris and his operations officer Lt Col Jim Kyle gave aircraft commanders final authority to determine the risk of any area thought to contain SA-2s. Some

AC-130 aircraft commanders did not change tactics at all, while others entered VR-4 with caution.

While Spectre 03's debriefing detailed both SA-2 engagements, 7th Air Force Intelligence discounted their report claiming there wasn't an SA-2 site in Laos nor could there be one established in such mountainous jungle terrain. The crew was absolutely certain they were engaged by SA-2 and a verbal battle ensued. Coincidentally, Spectre 12, commanded by Maj. Jenson, was working in a nearby sector reported being engaged by SAM, but did not receive any RHAW indications.[47]

That same night, Spectre 02 was diverted by *Moonbeam* to work a TIC for *Hammer 97*, located near route 914, ultimately spending their entire mission time working the TIC. Likewise, Spectre 06, commanded by Maj Herman Blahlock, was nearing the end of their mission when they were contacted by an OV-10 FAC to support a small friendly element trapped on a hill that was under attack by a much larger force. It was just another night in *Steel Tiger East* for AC-130 crews.

> "S-12 came into area 7 & 8, coordinated with them. They worked Delta 14, Delta 77 and Delta 82 while we remained south of Delta 12. Cleared this with Moonbeam. At 2107, had SAM indications and visual

47 It is likely Spectre 12 was engaged by a K-13 IR guided missile, which would not show up on RHAW gear. This was a secret PAVN program established specifically to shoot down an AC-130.

> sighting. Broken heading 210 moved to the west to Delta 77 (coordinated with S-12) informed Moonbeam of the SAM indications. At 2112, another SAM launch, broke again and due to aircraft stress RTB to CH 93." AC-130 Mission Report MSN NR 1125 2 Mar, 1971.

There was no shortage of crew chatter within the squadron. Some crewmen considered the SA-2 certain death, but intelligence analysts considered the SA-2 was slow to identify, track, and engage an AC-130. If the PAVN used standard Soviet doctrine when operating the SA-2, the average time from detection to launch was measured in minutes, not seconds, so the prevailing theory was an AC-130 crew would have time to avoid the threat if they reacted as soon as search radar activity was detected. If for some reason a missile was launched, it was unknown if defensive maneuvers or ECM pods had any effect at all. Because of these variable theories, some crewmen thought the AC-130 was a sitting duck, but others believed they were not likely to get hit. Maj Holley's crew suspected they were lured into a SAM trap since the convoy they were engaging when fired upon appeared too easy to find. Several days after Spectre 03's monumental escape, B-52 bombers and fighters supporting Lam Son 719 literally wiped the town of Tchepone off the map, destroying buildings, equipment, and personnel. Though it is unknown if the SA-2 site was destroyed in the process, no SA-2 indications were detected in the Tchepone area for the remainder of *Commando Hunt V.* More than a month after Spectre 03's encounter, 7AF intelligence finally admitted there was an SA-2 site near Tchepone.

Lam Son 719 ended on 25 March 1971. Efforts to sever the Ho Chi Minh Trail were largely unsuccessful and the ARVN suffered high personnel and equipment losses. The PAVN took heavy losses as well, but they managed to save Base Area 604. Nevertheless, critical intelligence concerning trail operations was discovered. During their sweep, ARVN forces repatriated Laotian forced laborers who worked the transportation network of the Ho Chi Minh Trail. Intelligence collected included personal observations from mechanics and trail workers. This human intelligence (HUMINT) was valuable information for *Spectre* crews since it explained why it seemed there was a never-ending supply of trucks and how damaged and destroyed trucks were handled.

> FM 7AF TAN SON NHUT AB RVN
> SUBJECT: TRUCKING, REPORTING NO.1
> THE FOLLOWING MESSAGE IS QUOTED
> FOR YOUR INFORMATION. "QUOTE"....
> SOURCE: RETURNEE, OMITTED
> A. TRUCKS, LAOS
>
> (1) SOURCE HAS NO KNOWLEDGE OF ANY TRUCK GRAVEYARDS IN LAOS/NVM: HOWEVER, HE SAW SOME TRUCKS THAT WERE 100% DESTROYED THAT HAD APPRERENTLY BEEN ABANDONED.

(2) SOURCE DOES NOT KNOW THE DISPOSITION OF DAMAGED TRUCKS HOWEVER, HE BELIEVES THEY ARE MOVED OFF THE ROAD AND HIDDEN UNTIL REPAIRS OR SALVAGE/CANNIBALIZATION OPERATIONS ARE COMPLETED.

(3) SOURCE THINKS DAMAGED TRUCKS ARE MOVED OFF THE ROAD BEFORE CANNIBALIZATION.

(4) THE MOVEMENT OF DAMAGED TRUCKS VARIES WITH DIFFERENT CIRCUMSTANCES AND CONDITIONS. SOME TRUCKS ARE LEFT FOR UP TO A MONTH BEFORE THEY ARE MOVED. DAMAGED TRUCKS ARE MOVED BY A 3-AXLED TRUCK EQUIPPED WITH A CRANE. EVERY CONVOY HAS AT LEAST ONE CRANE TRUCK.

(5) SOURCE KNOWS OF NO PRIORITY TO MOVE DAMAGED TRUCKS. DAMAGED TRUCKS ARE NORMALLY PULLED TO THE SIDE OF THE ROAD TO ALLOW THE REST OF THE CONVOY TO PROCEED. THE DRIVER OF THE TRUCK AND A CRANE TRUCK REMAIN BEHIND TO CAMOUFLAGE THE VEHICLE. THE TYPE OF CAMOUFLAGE DEPENDS ON THE TERRAIN. IF OPEN TERRAIN THE VEHICLE IS COVERED WITH DIRT. IF THE TRUCK IS UNDER THE

JUNGLE CANOPY IT IS COVERED WITH VEGETATION, HOWEVER CAMOUFLAGE IS NOT ATTEMPTED IN ALL CASES.

(6) IT IS VERY DIFFICULT FOR THE SOURCE TO ESTIMATE THE NUMBER OF TRUCKS OPERATING IN LAOS HOWEVER, THERE ARE PROBABLY BETWEEN 2 AND 3 THOUSAND BECAUSE THEY SUPPLY SO MANY TROOPS.

(7) SOURCE DOES NOT KNOW HOW MANY TRUCKS OPERATE ON ANY GIVEN NIGHT NOR DOES HE KNOW HOW MANY OPERATE IN A GIVEN BIHN TRAM'S OPERATING AREA. TRUCKS NEVER TRAVEL DURING DAYLIGHT HOURS IN LAOS AND VERY SELDOM DO THEY TRAVEL DURING THE DAY IN NVN. A MOONLIT NIGHT IS CONSIDERED TO BE THE BEST TIME TO TRAVEL.

(8) SOURCE IS NOT SURE OF THE PROCEDURES TO REPLACE DESTROYED TRUCKS, NOR DOES HE KNOW WHERE THEY COME FROM. SOURCE THINKS MOST TRUCKS COME FROM HANOI AND THE NUMBER OF TRUCKS REPLACED DEPENDS ON THE BINH TRAM COMMANDER (USUALLY A CAPT). A CLEVER COMMANDER ASKS FOR AND GETS MORE TRUCKS.

(9) A DIRECT HIT BY 40MM CANNON WILL DESTROY A TRUCK. SOURCE HAS BEEN TOLD BY TRUCK DRIVERS THAT THEY ARE VERY AFRAID OF THE AC-130S BECAUSE THEY ARE TOLD THEY ARE VERY ACCURATE AND SEEM TO STAY ABOVE THEM. SOURCE WAS ALSO TOLD THAT TRUCKS TRAVEL AT 50-100 METER INTERVALS AND THAT IS THE MIDDLE TRUCK IS ATTACKED THE TRUCKS IN FRONT WILL SPEED UP, AND THE TRUCKS IN THE REAR WILL TURN AROUND AND GO IN THE OTHER DIRECTION. DRIVERS HAVE ALSO TOLD SOURCE THAT THEY ARE NOT AFRAID OF HELICOPTERS OR JETS BECAUSE HELICOPTERS CANNOT FLY AT NIGHT AND JETS MUST CIRCLE AND PASS GIVING DRIVERS TIME TO LEAVE THE TRUCKS AND HIDE. SOURCE DOES NOT KNOW HOW MANY OR WHAT PERCENTAGE OF DAMAGED TRUCKS ARE REPAIRED AND PUT BACK IN SERVICE.

(10) SOURCE DID NOT SEE ANY CHICOM TRUCKS IN LAOS BUT HE HAS SEEN THEM USED IN NVN ON EASY ROADS (GOOD SURFACE, NOT TOO STEEP).

(11) MOST SPARE PARTS ARE SUPPLIED FROM HANOI.

(12) SOURCE DOES NOT KNOW IF CHICOM PARTS CAN BE USED ON SOVIET TRUCKS.

(13) IF A TRUCK TIRE DEVELOPS A SMALL HOLE, THE CONVOY CONTINUES ON, AND A CRANE TRUCK REMAINS WITH THE VEHICLE UNTIL IT IS REPAIRED.

(14) SOURCE ESTIMATES THERE ARE APPROXIMATELY 200 TRUCKS IN EACH BINH TRAM.

(15) SOURCE ESTIMATES THAT MOST CONVOYS CONSIST OF 50 TRUCKS.

(16) DRIVERS WILL ATTEMPT TO TRAVERSE B-52 STRIKE AREAS ONLY IF THEY FEEL THAT IT IS SAFE TO DO SO. OTHERWISE, THEY SIMPLY ABANDON THEIR VEHICLES. EACH DRIVER DECIDES ON WHAT HIS ACTION WILL BE.

(17) AC-130 GUNSHIP HAVE AFFECTED TRUCKING VERY MUCH. ALL DRIVERS FEAR THIS AIRCRAFT AND SEVERAL HAVE TOLD SOURCE THAT THEY ARE VERY ACCURATE, MAKE MANY PASSES AND HAVE SUSTAINED FIREPOWER.

(18) LAST YEAR, TRUCKS TRAVELED FASTER IN LARGER NUMBERS AND AT CLOSER INTERVALS (APPROX. 20 METERS) THAN NOW. TRUCKS ARE NOW TRAVELING IN SMALLER GROUPS (APPROX. 30)

AT SLOW SPEEDS AND AT GREATER INTERVALS (BETWEEN 50-100 METERS).

(19) DRIVERS PROCEDURES UNDER ATTACK VARY WITH TERRAIN. IF TRAVELING UP-HILL WHEN ATTACKED THE DRIVERS WILL ABANDON THE VEHICLES AND HIDE. IF ON LEVEL TERRAIN, APPROX HALF OF THE VEHICLES WILL SPEED UP WHILE THE OTHER HALF TURNS AROUND AND TRAVELS IN THE OPPOSITE DIRECTION.

(20) COORDINATION OF TRUCK MOVEMENT IS AFFECTED BY LAND LINE OF FIELD PHONES. EACH BIHN TRAM HAS 3 BARRIER STATIONS. THE BIHN TRAM INFORMS THE FIRST BARRIER STATION WHEN THE TRUCKS DEPARTED. THE FIRST BARRIER STATION INFORMS THE SECOND, ETC. EACH BARRIER STATION HAS THREE MEN RESPONSIBLE FOR TRAFFIC FLOW DATA TO THE BIHN TRAM.

(21) SOURCE STATES THAT AAA SITES ARE USUALLY LOCATED AT HIGH POINTS ALONG THE ROUTE AND ARE INFORMED BY THE BARRIER STATION WHEN TRUCKS WILL ARRIVE IN THE AREA BY FIELD PHONE.

(22) SOURCE STATES THAT THE DISTANCE TRAVELED IN A NIGHTS TRIP DEPENDS

> ON TERRAIN, BUT THE MAXIMUM DISTANCE IS 30 KILOMETERS BECAUSE TRUCKS GO FROM ONE STATION TO ANOTHER ROUND TRIP IN ONE NIGHT.

By April of 1971, the rainy season once again made roads impassable on the Ho Chi Minh Trail. AC-130 missions slowed to a trickle and crews launched found few trucks. PAVN gunners were also hindered by clouds and rain and the only battle damage received was on *Proud Warrior* when a 40mm projectile exploded while firing, causing minor fuselage and flap damage and injuring SSgt Adrian Richardson.

Commando Hunt V ended on 30 April 1971. Remarkably, the small fleet of fourteen AC-130 gunships built on antiquated and maintenance-prone C-130s was credited for more trucks destroyed or damaged than the combined efforts of hundreds of fighter and bomber aircraft. AC-130 BDA was indeed impressive; perhaps too impressive for some to believe. A few weeks before the end of *Commando Hunt V*, Lieutenant General Donald V. Bennett, Director of the Defense Intelligence Agency, stated that some top-level officials in Washington were questioning AC-130 truck kill totals. Air Staff passed a message outlining concerns to AC-130 commanders in Southeast Asia.[48]

> "AC-130 BDA is the hottest thing in the theater this moment. Seventh Air Force is really

48 Development of Fixed-Wing Gunships, 1962-1972, Ballard, P-168

> concerned about the validity of the BDA reported by the AC-130 gunships in their truck killing operation. They stated all aircraft for this hunting season indicates over 20,000 trucks destroyed or damaged to date, and if intelligence figures are correct, North Vietnam should be out of rolling stock. The trucks continue to roll however." MR Spec Ops Div, Dep Dir/Strike Forces, Dir/Ops, Apr 7, 1971.

While there was question as to the validity of truck kill numbers at 7th Air Force, there was no question as to the effectiveness of the AC-130 at *Group 559* Headquarters.

> "The number of trucks destroyed by enemy aircraft during the 1970-71 dry season rose to 4,000, of which the AC-130E (sic) by itself destroyed 2,432 trucks (60.8% of the total number of trucks destroyed by the enemy).[49] Our supply effort, conducted during a single season of the year and using "single supply route" could not respond adequately to our requirements. Our night supply operations encountered difficulties." Victory in Vietnam (PAVN History Archives) Mr. Merle L. Pribbenow, University Press, Kansas 2002.

49 The AC-130E was not employed until the 1971-1972 dry season. It appears most PAVN records indicate AC-130E as a general reference to all AC-130 models.

To answer the questions concerning AC-130's effectiveness, the 7th Air Force initiated a thorough analysis of BDA scoring criteria. It was universally agreed upon that analysis by the 16 SOS BDA Review and Evaluation Facility was accurate, but many factors could skew analysis to make hits appear to be more effective than they actually were. The immediate problem was traced to the "canned" kill criteria first levied by ASD, which did not consider physical variables crews typically encountered. Since Lt Col Jim Kyle was reassigned to 7th Air Force HQ, he assumed the duties of Gunship Division Chief to resolve AC-130 operational issues. 7th Air Force Commander Gen Lucius Clay was in a predicament with no easy answer regarding the accuracy of the reported BDA. Seeking clarification, Lt Col Harris and Maj Ed Holley's crew were summoned to Tan Son Nhut Air Base to meet with Gen Clay.

> "Our crew had flown a couple of special missions in which Gen Clay had personal interest. At those times, he made us feel as if he were working for us as much as we were working for him. That day at Tan Son Nhut was no different. I doubt that we told him anything he did not know or had not guessed, still, he listened. We said from our experience we believed that:
>
> -Crews relied on the single hit with a 40mm criterion mostly for expediency;

> -When time allowed, crews tried for burners and blowers;
>
> -Crew errors were honest mistakes" Truck Count, Air University Review, Henry Zeybel, Lt Col (Ret.)

A few weeks after their meeting with Gen Clay, Lt Col Harris and Maj Holley's crew, augmented by an additional Navigator, flew *Prometheus* to Tan Son Nhut to meet with Gen Clay a second time. This time, the General told them, "we might be working this problem backward, but the truth is all that matters. At the start of the dry season, we took ASD's word on what damage the weapons would do. Now I want to see it for myself."

Maj Holley's crew was then advised they would be flying a "scored mission" on a remote range north of Bien Hoa Air Base. Eight trucks were lined up on an S-shaped road, with six spaced thirty–forty feet apart, staggered left and right of the center of the road. Two more were over one hundred feet behind the others. Gen Clay would personally direct the crew to fire specific ordnance on each target while viewing from a bunker about one thousand meters away. The entire event was carefully planned and orchestrated by Gen Clay's staff. Maj Holley's crew did not know specifics, but several trucks would have running engines, one was shut down but still warm, and the rest were cold. Some trucks would be loaded with fuel barrels and some with non-flammable cargo.

When Maj Holley's crew arrived over the convoy, they were asked to report how many trucks had running or warm engines and they identified four total. Gen Clay verified they were correct, then directed the crew to fire on the lead vehicle. Maj Holley directed the LLLTV as the primary sensor and requested gunners arm the 40mm guns. A few minutes later, a sensor operator centered his crosshairs on the hood of the lead truck and Maj Holley began to fire.

> "We opened fire with a vengeance, not using normal rhythm; instead pouring out a long stream of rounds. The first shell exploded in a roadside ditch. "five low," was the reflexive comment. The stream of rounds walked out of the ditch, arced across the road, and smacked all over the lead truck." Truck Count, Air University Review, **Lt Col (Ret.) Henry Zeybel**

For the remainder of the evaluation, Gen Clay directed the gunship crew to fire on specific targets with 20mm and 40mm using both sensors while scoring every shot via the usual method. The crew was advised to stop firing only when they believed they had destroyed the truck. Holley's crew hit the lead truck numerous times, but it did not burn. Trucks loaded with fuel barrels burned spectacularly when hit. The last two vehicles were hit numerous times with 20mm HEI but did not burn. When the test was complete, the crew landed on Bien Hoa and were transported to the range to review BDA results first-hand.

They found 20mm HEI was ineffective when fired from above 7,500 ft. altitude since most projectiles did not explode. Hits on the hood or cab caused sheet metal damage but rarely penetrated into the engine compartment. Some direct hits on the cab, hood, or fenders dented sheet metal and bounced off without exploding. If a projectile hit a hard surface, it would explode, but the surface had to be harder than typical soil on the trails in Laos. Fragments caused window glass to crack, but little damage was done to sheet metal. In all, trucks hit by 20mm appeared to be easily repairable. The effects of 40mm HEI were better than 20mm, but not significantly. A direct hit of a two-pound 40mm projectile could kill a truck, but only if it hit a vital component such as the engine or fuel tank. Direct 40mm hits on the bed of a truck loaded with supplies may damage the supplies, but fragments and blast were absorbed by the cargo, leaving the truck relatively undamaged. From a gunship visual sensor perspective, what appeared to be a spectacular direct hit when viewed on the IR or LLLTV may or may not have actually caused significant damage.

Gen Clay, Lt Col Harris, and Lt Col Kyle were satisfied with the results of the test which verified AC-130 crews were reporting BDA according to ASD criteria without exaggeration. At the same time, ASD kill criteria was discovered faulty and steps were taken to change kill criteria to reflect observed effects. The test helped explain why it seemed the PAVN had a never-ending supply of trucks, when in fact a sizeable percentage of trucks claimed as destroyed were actually repaired and returned to service.

> "In May of 1971, I was transferred by Col Harris to 7th AF to assume duties as the Gunship Division Chief. While there, a BDA validation project was directed by Gen Clay. We acquired several salvaged trucks and positioned them on the Bien Hoa firing range ... an AC-130A attacked them from various altitudes using 20 and 40MM. It became clear from BDA results that criteria were wrong and 7AF published new rules as well as adjusted truck kill totals." **Col (Ret.) Jim Kyle,** Pilot

Since there was no question a truck blown apart by explosive cargo or engulfed in flames wouldn't be destroyed, the BDA criteria was amended to reflect a truck kill would only be credited if it burned or exploded. Trucks that were hit directly without burning or exploding were considered only damaged. As a general rule, the new criteria meant more time and shots needed per engagement to kill a truck.

The 16 SOS spent the summer months on strip alert for on-call TIC fire support tasking. Many times, the rain was so heavy crews were unable to take off, or even taxi. At least some missions were flown into northern Laos where the weather was only slightly better than in the south. A total of eight AC-130A gunships remained on station during the rainy season while the remaining fleet, including The

Surprise Package aircraft (*Thor*), returned to CONUS for rework and modifications focused on offensive and defensive improvements. To counter the SA-2 threat, ASD engineers tested and installed ALQ-87 ECM jamming pods to provide at least some level of protection against the SA-2 when combined with defensive maneuvers and manually dispensed chaff packets.

By the fall of 1971, the 16 SOS had doubled in size as newly trained crews and maintainers arrived while personnel reaching the end of their one-year tour departed. Three of the first built AC-130A gunships and the first two production AC-130E gunships were assigned to the 415 SOTS at Hurlburt Field, FL to train new crews. The extreme size and turnover of personnel made it nearly impossible to train replacement personnel via mentorship, so experienced aircrews at Ubon developed and published the first-ever AC-130 tactics manual on 1 October 1971. This living document was the fundamental standardization baseline for all gunship tactics, and crews were encouraged to contribute lessons-learned into the basic manual as enemy tactics changed. A "hard crew" method of scheduling was implemented whereas crewmen would fly with the same crew at all times. This concept contributed to better crew coordination and was thought to improve esprit de corps. The annual "passing of the torch" began with crewmen about to end their tour briefing newly arrived crewmen. These events were timed to begin just before the annual dry season arrived when weather in *Barrel Roll* was marginal but good enough to familiarize new crews with the Laos VR sectors. By early October 1971, 144 combat missions had been flown. Though gunship crews located few trucks, they laser-designated road intersections and suspected storage areas for the F-4s to drop laser-guided bombs on pinpoint targets. Overall, these missions helped new crews prepare for heavier truck traffic expected in the coming months.

Beginning in mid-October 1971, AC-130As *Ghost Rider*, *Night Stalker*, *The Exterminator*, and *Thor* returned to Ubon. *Thor* was equipped with experimental equipment including a Gated Laser Intensifier for Night Television (GLINT) which produced laser illumination for the LLLTV sensor that could not be detected with the naked eye. The GLINT greatly improved LLLTV imagery on nights void of starlight or moonlight and was much better than the 2KW light. Another laser illuminator called the "green weenie" produced a beam of overt high-intensity green light that was used to designate ground targets for conventional bombing by fighter escorts.[50] Both lasers and a newly developed helmet sight were to undergo trials during combat operations. Gunship technology improvements continued on the rest of the fleet as well. In October, a 30 man civilian team from Ling-Temco-Vought (LTV) arrived at Ubon to update the eight remaining *Pave Pronto* aircraft with the AN/AAD-7 IR sensor, electronic countermeasure (ECM) pods, the LTN-51 inertial navigation/targeting system, and 40mm ammunition storage racks. Just about the time *Commando Hunt VII* was

50 The green weenie was universally despised by most crews for use over the trail since it illuminated the aircraft's position. It was not often used for combat operations but was sometimes used to trick AAA gunners to hastily abandon their guns for fear a bomb was inbound.

about to begin, Typhoon Hester flooded major portions of the Ho Chi Minh Trail and halted PAVN truck traffic on the trail.

By the third week of October 1971, a total of 687 personnel were assigned to the 16 SOS. Billeting arrangements and squadron buildings were improved. Amongst 185 newly arrived personnel who began indoctrination training, there were veterans from the previous season who had either worked a deal with Air Force Manning & Personnel Center to follow or found a way to work the system to return within a few months of departure. Of the most helpful at AFM&PC was MSgt Brian "Morrie" Morris, known as "The Spectre Protector." MSgt Morris knew the system well and many *Spectres* returned for second and even third tours because of his insight and knowledge of the assignment system.

As the squadron grew, traditions were established, usually at the expense of new arrivals. The unit mascot was a live Turkey named Ken, who was led around on a leash by the lowest ranking officer assigned. Crewmen had pets also, including a monkey and a Gibbon Ape that made regular appearances at squadron gatherings. Morale was high, especially among gunners who adopted the squadron's unofficial theme song "*Ghost Riders*", which was loaded on every jukebox on and off base. Though all crewmen shared the same level of camaraderie while flying missions, gunners took the concept of self-esteem and unit pride to the extreme. Gunners established Rules Of Engagement (ROE) requiring everyone to stand whenever any version of the song was played and anyone who refused to stand risked a physical altercation. Often, a group of "well liquored" *Spectres* would form a large circle upwards of twenty people, place their arms on each other's shoulders, and perform a dance akin to a Russian "khorovod" folk dance. Many times, the dance circle produced bruises, battered shoulders, or a bloody nose as a result of a "circle crash" by falling face-first onto the floor. At the end of the song, everyone yelled as loud as they could, "if you ain't *Spectre*, you ain't shit." Of course, there never was a shortage of alcohol at a *Spectre* gathering, but aside from a few hangovers the following morning, crewmen were ready to saddle up to hunt trucks on the Ho Chi Minh Trail once the sun went down.

Another chapter in *Spectre* operations began on 25 October 1971 when AC-130E, aircraft 69-6569 (*Excalibur*), arrived at Ubon. Six days later, *Operation Commando Hunt VII* began and the 16 SOS began flying an average of fifteen missions per night. *Excalibur* flew its first combat mission on 2 November 1971. Since it could loiter longer and had more powerful engines, mission planners sent it into *Barrel Roll* frequently. Just as the first AC-130E was cutting its teeth in combat, the Gunship Program Office approached PACAF with a request to modify an AC-130E in the modification line with a large caliber gun. Considering the ongoing level of effort to interdict heavy truck traffic on the trail and the need for as many AC-130s as possible for combat operations, the 7th Air Force took a dim view of their proposal. Fearing the experiment would take valuable time away from the *Commando Hunt VII* interdiction campaign, the 7th Air Force denied the Gunship Program Office's request.[51]

Commando Hunt VII proved to be the busiest campaign yet, but missions settled into a routine pattern for Capt Charles Baertl who started his tour in early 1971 as *Spectre* pilot. On 8 November, he was the aircraft commander of Spectre 03, flying *Raid-Kills 'Em Dead* when the aircraft sustained slight battle damage. A few nights later, he and his crew destroyed 8 trucks and damaged 1 in *Steel Tiger East* VR 2 and 3 before developing fire-control computer problems. At the time, Capt Baertl had flown upwards of seventy missions into Laos, hunting trucks as a matter of routine. Then, on the night of 13 November 1971, merely two months before the end of his tour, he was directed to take the place of another aircraft commander who was sick and unable to fly. At the time, he had no way of knowing the mission he was about to fly would result in a true legend of aviation.

His mission began just like every other as his crew arrived at the squadron building for their pre-mission brief. Since Capt Baertl was standing in for the sick pilot, he was not flying with his regular crew, but the crew was highly experienced nevertheless. After an uneventful take-off, *Prometheus*, (call sign Spectre 21) entered *Steel Tiger East* VR-5 for a routine armed reconnaissance mission. Approximately 20 seconds after rolling into firing orbit, a

51 The second AC-130E (69-6571) was delivered to Ubon without delay on 6 Nov, and the third (69-6572) was delivered on 10 Nov.

muffled explosion and bright flash were observed near #3 and #4 engines.

> "We were on our first orbit. The BC had picked up a stationary ignition signal from what was thought to be a truck. I was scanning for AAA fire, and when I turned my head to the 3 o'clock position, two to three red flashes came from the right low side of the aircraft. The aircraft bounced and shuddered like it had hit a pothole. Prometheus began a slight nose down descent that was felt by all aboard." **SMSgt (Ret.) Bill Patterson, IO.**

The aircraft began an uncontrollable roll and was losing altitude while Capt Baertl and his copilot Capt Dennis Carlson fought to regain control. Suspecting they had been hit, Capt Baertl directed the crew to prepare for bailout while navigator, Capt Garry Witt called out altitude information and a heading to return to Ubon, which was the closest airfield. At the same time, crew members hurriedly donned parachutes and waited for the order to bail out. While the cockpit crew wrestled to regain control, two gunners rushed to assist the right scanner, who was stunned and cut up a little, but otherwise unhurt. Thankfully, after losing 2,500 ft. of altitude, the pilots regained control of the aircraft and leveled off. Still unsure what had happened, everyone was relieved that bailout was less likely, but it was a mystery as to why the aircraft required excessive rudder input to keep on course. Finally, a gunner looking out the right scanners hatch keyed his microphone and said “both right engines are gone.” Flight engineer, MSgt Larry Stauty, could not believe what he was hearing since all engine gauges seemed normal, so he jumped out of his seat and ran to the right scanners to have look. To his amazement, both #3 and #4 propellers were missing and flames were coming out of the #3 engine. MSgt Stauty quickly returned to the flight deck, activated the engine fire suppression system, and shut down both engines. Though Spectre 21’s immediate emergency was resolved, the aircraft was too heavy to maintain altitude on only #1 and #2 engines. For the next few minutes, the pilots tried everything they could to gain altitude, but nothing worked. Capt Witt realized more altitude was needed to clear the mountain peaks en route to Ubon, but *Prometheus* could not climb, even at maximum effort. Capt Baertl turned control of the aircraft over to Capt Carlson while he and MSgt Stauty discussed options to lighten the aircraft. The first option in most aircraft was to dump fuel, but the C-130A was not capable of dumping fuel. As they continued to discuss options, Capt Baertl was considering bailout as their only remaining option, but the thought of spending the night in the Laotian jungle was not appealing in the least bit. As discussions continued, SSgt Patterson keyed his microphone and said “how about we throw out the ammunition?” Capt Baertl agreed with SSgt Patterson, then ordered jettison of all ammunition. During the next five minutes, gunners and IO’s jettisoned over seven thousand lbs. of ammunition and equipment.

> "I was in the process of getting my annual flight evaluation by SSgt Tony DeNoi when the props came off. We began to jettison everything that wasn't bolted down from the aircraft. Tool kits, ammo cans, you name it, all gone to lighten the aircraft. At one point, we had a long belt of 20mm ammunition trailing the aircraft in the slipstream and static electricity started igniting electric primers causing flashing and popping noises behind us. It was just one bizarre thing after another that night." **SMSgt (Ret.) Bill Patterson, IO.**

Now three and a half tons lighter, *Prometheus* was able to climb and maintain just enough altitude to clear the mountains. Capt Witt plotted a course to fly between the lowest part of the terrain saddle, but everyone knew it was going to be close. After passing over the mountains by only a couple hundred feet, the entire crew breathed a sigh of relief, but the drama was not over yet. About 12 miles from the Ubon runway, #2 engine instruments indicated turbine failure with imminent power loss. SSgt Patterson kept his eye on the engine which was glowing red-hot and throwing sparks. Knowing the aircraft would quickly stall and crash with only one operational engine, Capt's Baertl and Carlson and MSgt Stauty carefully manipulated power, trim, and flight controls and flaps until they slammed down on the runway at a speed of about 140 knots as if they were landing on an aircraft carrier. Stopping the aircraft was also challenging since braking was typically done by main landing gear brakes and reversing of the props on all four engines, which was impossible. Capt Baertl knew brakes alone may not stop the aircraft before running off the end of the runway, so he gingerly reversed the left side props. This caused *Prometheus* to veer to the left side of the runway until straddling the blue runway clearance lights until Capt Baertl brought the aircraft back onto centerline with nose wheel steering. Finally, Spectre 21 screeched to a halt on the taxiway as smoke poured off red-hot main landing gear brakes. Without hesitation, Capt Baertl rang the alarm bells and everyone egressed from the aircraft. Fortunately, if not miraculously, there were no serious injuries during the entire ordeal. The crew debriefed their mission, filled out statements, and went directly to the Officers Club where they stayed until sunrise.

> "When we arrived at the O-Club, someone gave me a bottle of Jim Beam and I did my best to drink most of it. We joked it was "just another routine two-engine out landing" but nothing could be further from the truth. When I look back at that mission many years later, I am amazed we actually made it back. Though many people congratulated me for excellent piloting skills, the actions of rest of the crew were equally impressive. Of particular note, I credit SSgt Patterson

> for saving the aircraft since he recommended jettison of ammunition and equipment. Had we not done that, we would have either had to bail out or, since we were unable to climb, crashed into the mountain top on our way back to Ubon. It's been said that flying military aircraft consists of hours and hours of sheer boredom, interrupted by minutes of sheer panic. Fortunately, I was so consumed with the task at hand that I was not afraid, though I probably should have been. Of all the missions I have even flown during my career in the USAF, that mission stands out above all others. **Lt Col (Ret.) Charles "Chuck" Baertl,** pilot.

When the sun came up, an inspection of the aircraft revealed heavy damage to both engines, but there were only a few holes in the right wing and fuselage. Most of the crewmen returned to the aircraft to take a look and figure out what happened. Regardless of effort, it was difficult to determine a cause and it appeared the loss of the second prop was a result of being hit by the first. Even after a close inspection, some believed AAA was the culprit while others surmised prop blade loss or gearbox disintegration. Though the actual cause was never determined, Capt Baertl and his crew demonstrated an incredible feat of airmanship that night, a feat most thought was impossible.

Damage to Spectre 21 that occurred on the night of 13 Nov, 1971. The actual cause of the loss of #3 & #4 propellers was never proven. (USAF Photo)

The crew of Spectre 21 (Crew of the Month—November, 1971) in front of the 16 SOS squadron building. Front row, L to R, Sgt Wolf, SSgt Patterson, SSgt Denoi, Capt Carlson, Maj McDonald, Capt Baertl. Second Row, Sgt Frost, Sgt Bedard, TSgt Warner, Maj Mackrell, Capt Witt, Lt Derosa. Rear row, Capt Sparkman, Maj Miller, MSgt Stauty, Capt Kelley. (USAF Photo)

The same night, *Night Stalker* (call sign Spectre 05), commanded by Capt Hobgood, searched for targets in *Steel Tiger East* VR 4. After tracking a truck into a camouflaged transshipment area, they opened fire, ultimately destroying 8 trucks and starting numerous secondary fires. The weather was not cooperative, so they were unable to bring their fighter escorts onto the target before the end of their mission.

Though most AC-130 missions were sent into the Ho Chi Minh Trail VR sectors, four significant TIC missions into South Vietnam were flown on 19 November 1971. Weather was bad while supporting friendly forces, so the crew relied heavily on their APQ-150 radar to identify friendly beacon locations. Throughout the night, four AC-130s tag-teamed to provide fire support for *Zulu Charlie* to repel aggressors, sometimes as close as 50 meters from friendly forces. An AC-130 remained overhead until 0630 when the ground team was successfully extracted by helicopter.

> "The ground commander gave us a 30-50 meter firing distance. We couldn't see the ground, friendly position or target area. The only sensor that was usable was the APQ-150 which had a good fix on the ground commander's beacon. We determined this to be a very high risk TIC and tried to talk him out of it, but he said they were hunkered down in a bunker and to fire away! We could hear the rounds hitting over the ground radio.

> There was no margin for error." **Maj (Ret.) Hank Rother,** EWO.

An AC-130E milestone occurred on 22 November 1971 when *Excalibur* (call sign Spectre 05), commanded by Capt King, located twenty-five trucks, destroyed nine, and damaged sixteen in *Barrel Roll.* This was the first significant truck kill mission for an AC-130E.

December 1971 brought in good weather, but trucks were getting more difficult to find, even with the Black Crow sensor. Intelligence reports surmised the PAVN had finally developed effective countermeasures to avoid detection. Target acquisition was difficult also since the PAVN had drastically changed tactics from the past season. Their early warning system was more effective and the large, compact convoys that were so common during *Commando Hunt V* were no longer found. Instead, each truck had to be located individually and most were well spaced and camouflaged. Frequently, trucks were found on side roads or newly built trails adjacent to the main trail to avoid detection, but visual sensors remained effective as IR and LLLTV sensor operators located hundreds of trucks every night. AAA fire was heavier than any previous year and F-4 escorts were not always available, so AC-130 vulnerability to AAA was a major concern. When height finder radar signals were detected near Tchepone, the emergence of radar-directed 57mm guns was likely. Since there was no possibility either model of AC-130 could operate high enough to escape the effective range of radar 57mm AAA, crews could not mitigate the threat. Previously, 57mm guns were optically aimed with tracers and usually fired in four-round bursts, with only the first round being a tracer. Just like tactics to avoid 37mm AAA, AC-130 crews relied on tracers to detect hostile fire to avoid being hit. If tracers were not used, crews were unable to detect or avoid 57mm AAA fire. Obviously, this was a major concern for all AC-130 operations.

Aside from 57mm, 37mm AAA was heavy in nearly all areas in *Steel Tiger East* and *Barrel Roll*. Sometimes AAA was so overwhelming that crews were forced to pull off target to work a different area. Such was the case on 10 December 1971 when *Excalibur* (call sign Spectre 02), commanded by Capt Rowland, received 1,409 rounds of 37mm AAA in the vicinity of Ban Ban in *Barrel Roll*. Though the crew managed to destroy three trucks and damage three more, they spent more time evading AAA and vectoring fighters onto gun sites than firing on trucks.

The last big truck kill mission of 1971 occurred on the night of 30 December as Spectre 14 neared the end of their mission. On that evening, *Nightstalker*, commanded by Capt Vaughn, located a string of fifteen northbound trucks. Their 20mm guns were inoperative, so they used 40mm guns to destroy five trucks and damage three. When all ammunition was expended, they FAC'd for Redbird 02, a B-57 who made five passes, then two F-4s, call sign Gunfighter 01 and Gunfighter 02, to bomb the remaining trucks.

"They would be on the perch waiting for a gun to come up, attack quickly and silence

> or destroy it. I was on missions when they would call "Winchester" and move away from us and wait. When a gun would come up they would dive in to attack the site, those guys had balls." **SMSgt (Ret.) Frank Hanna,** Gunner.

Data collected from AC-130 missions during November and December 1971 credited AC-130 crews with 782 trucks destroyed and 1,141 damaged, three times greater than the same period one year earlier. These increases were commendable, but TIC tasking and marginal weather affected the overall truck kill potential, which could have been substantially higher.

By January 1972, all but two AC-130s were equipped with chaff dispensers in the LAU-74 flare launcher. This system, coupled with gunship-specific SAM evasion maneuvers gave limited protection against the SA-2. Furthermore, if an LAU-74 was not operational or available, the previously established tactic of using manually dispensed chaff packets would be employed.

The weather for the first half of January 1972 was spotty. Crews were spending more time engaging trucks to meet the new truck kill criteria, increasing exposure to enemy AAA. To avoid exposure to AC-130 attacks, *Group 559* commanders shifted operations to the early morning hours to avoid an attack. *Spectre* began daylight operations teamed with B-57 *Redbird* bombers under *Ubon Plan Charlie*. Missions were flown at 15,000 ft., to locate trucks with the IR and

LLLTV, then laser designate targets for the *Redbirds*. After about a dozen sorties, few trucks were detected and 57mm AAA gunners could easily acquire an AC-130, which made for unpleasant surprises. During one of these missions on 12 January 1972, Spectre 13 was flying at 13,500 ft. when hit by a burst of 57mm AAA. The IO was wounded and flares in the LAU-74 launcher were set on fire. Fortunately, the IO was able to jettison the burning launcher before smoke and fire overwhelmed him. Since daylight operations proved to be high-risk/low gain, day missions were discontinued after less than a dozen sorties. The entire squadron again shifted their focus to nighttime operations escorted by *Redbirds* as a hunter/killer team. These missions proved effective against tanks, armored vehicles, and heavy artillery, which were typically invulnerable to AC-130 gunfire.

AAA wasn't the only threat in the congested VR sectors of *Steel Tiger East*. Because of high levels of air traffic in small VR-sectors, there was substantial risk of in-flight collision with other aircraft or ordnance being dropped through AC-130 orbits. On several occasions, B-52 *Arc Light* strikes fell through AC-130 orbits. After these near-tragic incidents, command and control of *Steel Tiger East* VR Sectors 4A, 4B, and 5 were transferred to *Task Force Alpha* (call sign *Warlock*) located at Nakhon Phanom, Thailand. The primary objective was to provide timely traffic advisories to prevent AC-130 collision potential, but in an ironic twist, *Warlock* sometimes sent two AC-130s into the same area at the same altitude, increasing the risk of collision. With no other options available, gunship crews took it upon themselves to de-conflict altitudes via real-time radio relay.

By early 1972, *Spectre* crews had nearly mastered defensive AAA tactics while wreaking havoc on supply trucks on the Ho Chi Minh Trail. At the same time, General Nguyen and his staff exerted tremendous pressure on the PAVN air defense network to shoot down a gunship. When 7th Air Force Intelligence intercepted a PAVN message from General Nguyen offering a reward for any PAVN gun crew who shot down an AC-130, Spectre crewmen devised a wholly unauthorized project appropriately titled "Pave Pig." Their objective was to parody General Nguyen's reward by dropping a live pig via parachute as a reward for substandard marksmanship. The mission went off without a hitch as Spectre deliberately mocked General Nguyen, his AAA gunners, and *Group 559* drivers.

> "Intel informed us the NVA was going to reward any gun site that managed to shoot down a Spectre Gunship. Not to be outdone, Jim Spiers crew made a wooden cage from a MK-6 box. They went downtown, bought a live piglet and brought it back to base. They put the piglet, a can of beer, a carton of cigarettes and a Spectre lighter inside. They taped a flashlight in one corner and a couple of day time survival flares on other corners and attached a flare chute to the box. They put a note in the box written in Vietnamese

> thanking them for substandard marksmanship, loaded it on the plane, and took off. When over their working area in Steel Tiger, the gunners threw the Piglet out over the first gun site that opened fire." **MSgt (Ret.) Dave Burns,** gunner

Aside from multiple limiting factors, the weather cleared for the last two weeks of January 1972. During an eight-day period, the 16 SOS averaged 100 trucks destroyed/damaged per night, but *Group 559* continually changed tactics to fend off AC-130 attacks. On the night of 18 January 1972, Spectre 21 was working a target in VR 4B when a cluster of five or six artillery flares burst about 2,000 ft. above their orbit. The crew broke away from the flares, then continued their attack, ultimately destroying five trucks and damaging a sixth. During their mission, they took 666 rounds of 37mm, 159 of them accurate. Flare launches were noted by other crews during follow-on missions, and it seemed a new tactic was emerging. In response, the squadron evaluated potential vulnerability at Ubon by dispensing flares from a high aircraft to illuminate the low aircraft. They found virtually no indication the aircraft was illuminated by flares and considered flares more of a harassment element than a threat.

Unknown to US Intelligence in late January 1972, PAVN engineers had completed a secret trail cut through triple canopy jungle for the sole purpose of preventing air attacks. Dubbed the *"K-Trail"* by North Vietnam, about 70 percent of its total 700-kilometer length was concealed under triple canopy jungle. It took almost one year to complete and enabled daylight travel of convoys of thirty-six trucks, which were nearly undetectable by air for almost the entire length of the trail. To maintain tactical deception, most of *Group 559*'s routine operations continued, but in smaller convoys on well-known established trails. This was an intentional plan to fool U.S. Intelligence into thinking the interdiction campaign was more successful than it actually was. Meanwhile, large convoys of supplies and combat divisions flowed to the south via the *K-Trail*.[52] Because of this major change in strategy, AC-130 crews found it more difficult to find convoys on familiar trails, but truck parks were still found. On 26 January 1972, a well-coordinated mission occurred in *Steel Tiger East* VR 4B when Spectre 12 and Spectre 18 joined their escorts Jackal 01, 02, and Olds 01. After *Spectre 12* located a major truck park near Tchepone, they began to engage targets one by one while fighters suppressed active 37mm AAA sites. In what turned out to be a six-hour attack by a tag-team of Spectre 12, Spectre 18, and multiple fighters, all truck activity in sector VR-4B halted. When the attack was complete, the truck park was in shambles with multiple fires and secondary explosions ongoing. Damage was assessed as fifteen trucks destroyed, thirty-seven damaged and *Group 559* had lost a major hub in their network.

Though heavy traffic was flowing down the *K-Trail*, the original trail network was still operating and AC-130

52 The existence of the "secret trail" was not discovered by US Intelligence until after the war ended.

crews continued to find and destroy small convoys, hidden Binh Trams, and truck parks resulting in significant losses. It was obvious to General Nguyen's staff something drastic was needed to stop AC-130 attacks. Their final decision was ordering the PAVN 67th Missile Battalion to establish a secret SA-2 missile site deep into the jungle of Laos to ambush an AC-130. The area chosen was about twenty-one miles southwest of Tchepone called "*May Huc*." It was a crude spot bulldozed into the jungle to overlook a known trail entrance between Hills 646 and 159, where gunships routinely passed through to avoid climbing over high terrain. The plan to transport the complicated SA-2 system was physically and logistically challenging since the system needed to be covertly transported piece by piece to the site, then reassembled on-site without being discovered.

> "To ensure secrecy and safety, the battalion worked out a very detailed and carefully controlled plan for the move. The unit was divided into individual sections for the move. The missile guidance van, the transceiver van, and the fire-control computer van; three trucks carrying antennas; three electrical generators and the electrical distribution vehicle; and the K-67 crane formed one section. The P-12 radar and the communications vehicle with the R-405 radio relay formed another section. 2nd Battery was divided into two sections, each section

> consisting of two missile launchers and three missile transport vehicles." PEOPLE'S ARMY OF VIETNAM AIR DEFENSE SERVICE People's Army Publishing House, Hanoi, 1994. English translation by Merle L. Pribbenow CIA Operations Officer (Ret.)

The plan was especially risky since the 67th Missile Battalion needed to transport multiple loads of outsized equipment on Route 12, directly through an area frequently hunted by *Spectre* crews. To avoid detection, transport vehicles were camouflaged to the maximum extent possible and all work was conducted in the dark without lighting of any kind. Road guards were posted to guide drivers towards the site and to prevent them from getting lost. The last few kilometers near the intersection of Routes 138 and 9 had been cleared by bulldozers. The launch site was small, narrow, and suitable for only two missile launchers, the *Fan Song* Radar, and support equipment. Because of terrain and covert mode of transport, trail workers encountered difficulties moving the cavalcade of equipment and supplies needed to establish a workable SA-2 site. Dozens of trail workers were committed to masking the presence of activity on and near the secret launch site. Not only were components of the system camouflaged, but the trails leading to the site were concealed by hundreds of potted plants in bamboo baskets that could be moved out of the way when required. From the air, there was no overt sign anything was there, but it would take some time to establish an operational SA-2 site.

In mid-February, Spectre 18 was flying a routine mission in *Steel Tiger East* VR 4 when they detected an India band radar lock-on and accurate 57mm AAA fire. After several more engagements and close calls, Spectre 18 departed the area. When the crew debriefed their mission, everyone acknowledged the fact that radar-directed 57mm was a serious threat escalation, but missions into *Steel Tiger East* VR 4 continued. Most believed the only way to mitigate the 57mm threat was to operate at higher altitudes, but climbing higher incrementally degraded the effectiveness of the 40mm gun. Fortunately, ASD was already working on a plan to retrofit a 105mm artillery gun onto an AC-130 to increase range and lethality. When the initial idea was brainstormed in mid-1971, there were some who thought ASD engineers needed to be locked in a padded cell. Regardless of opposing opinions, there was merit from an engineering perspective. Under the project name *Pave Aegis,* an M102 Howitzer was installed on aircraft 54-1626. With promising results in hand, *Pave Aegis* flight tests were completed on an AC-130E in January 1972 at Hurlburt Field FL.

When the Gunship Program Office briefed Secretary of the Air Force Robert Seamans Jr., the 105mm gun was touted as out-ranging 57mm AAA and capable of firing an accurate high explosive (HE) projectile sixteen times heavier than a 40mm projectile. ASD engineers determined addition of a 105mm gun would not only increase stand-off range but would increase lethality as well, then requested one AC-130E be diverted from the production line for *Pave Aegis* modification. Secretary Seamans viewed the proposal favorably, but General John D. Lavelle, CINCPACAF, considered the proposal risky. He was unconvinced there was a need for the 105mm gun, nor did he want to lose a valuable aircraft to an unproven weapon system during the height of *Commando Hunt VII*. Fortunately, CSAF Gen John Ryan and Secretary Seamans felt the advantages of flexibility, survivability, and effectiveness warranted a go ahead and directed PACAF to commit to a thirty-day combat evaluation.[53] The gun was shipped to Ubon via C-141, accompanied by an engineering team and an initial cadre of crewmen and maintainers. On 17 February, 69-6574's (*Iron Maiden*) APQ-150 radar was removed to make room for the cannon, which protruded from the left paratroop door. The gun mount was equipped with manual hydraulic hand pumps to move from the stow position to azimuth and depression settings needed for firing. Two ammunition racks capable of storing a total of 96 rounds were added to hold 42lb 105mm cartridges. The mount, ammunition, and associated equipment weighed over 7,300 lbs. and was the largest gun ever installed on any aircraft.

The *Pave Aegis* combat evaluation commenced on 22 February 1972, led by 16 SOS project officer Lt Col Thomas Debnam, 16 SOS standardization evaluation gunners, and crewmen who participated in CONUS testing. This crew, commanded by Capt Waylon Fulk, was assigned to the *Pave Aegis* aircraft for the duration of the combat evaluation. After an uneventful take-off, they flew directly to

53 Project Checo, Pave Aegis Weapon System, 30 July 1973, P-12

a permissive area in northern Cambodia for a live-fire to acquaint the crew and develop coordination procedures. Equipment problems and ammunition misfires caused the crew to return to Ubon for maintenance. Other problems were noted, but the crew adjusted without difficulty. A successful training mission was flown on 23 February, and the next night Capt Fulk's crew flew the first *Pave Aegis* combat mission into *Steel Tiger East*. During their mission at 8,000 ft. altitude, they destroyed three trucks and damaged four with the 105mm gun, and destroyed/damaged eight more with their 40mm gun.

The 105mm gun was tested on an AC-130E at Hurlburt prior to being sent to Ubon for combat evaluations. (USAF Photo)

In mid-February, a *Spectre* crew was flying a routine armed reconnaissance mission in *Steel Tiger East* VR-4 when Capt Gerry Stiles, the LLLTV operator, spotted what appeared to be shaded headlights about nine miles distant. Strangely, the lights appeared to be traveling west into Laos and not south or east as trucks normally did. At first, the crew joked the driver might be lost, but when the vehicle's lights went out, they decided to investigate. Arriving overhead, they found a narrow trail that led them directly to a small Jeep-like vehicle intermittently flashing its lights as if signaling. Furthermore, when Capt Stiles searched up and down a narrow trail cut through high trees, he realized the vehicle was leading a convoy of about a half-dozen vehicles. Included in the convoy was an odd-looking semi-tractor trailer loaded with what appeared to be a large missile. Using standard AC-130 tactics, they engaged the convoy with 40mm HEI until a large secondary explosion occurred. As the crew rolled out, they were certain they had just destroyed an SA-2 in transit. After landing, 16 SOS intelligence analysts reviewed the BDA videotape and deemed the target was indeed an SA-2 but wondered why it was found so far into the Laotian jungle.

> "I moved my aimpoint to the front of the object to work over the remaining vehicles and shot the tractor which started flaming- it wasn't going anywhere. Next, I moved to the long body of the missile-like object and scored several hits on it.- all too little or no

> avail for we saw not flame or smoke. Now, suspecting it was an unfueled SAM, I moved to the front, pointy end, of the long object and continued to fire the 40s. We scored several near-misses on the front portion when "kaboom" we hit it and got a massive secondary explosion which proved to be the SAM's solid explosive warhead." **Maj (Ret.) Gerald "Gerry" Stiles.**

When 8th TFW Intelligence reported an SA-2 was found in Laos, Seventh Air Force Commander General John Lavelle and his staff scoffed at the claim as if it were a UFO sighting. What followed was a profound display of arrogance as 7th Air Force Intelligence took no action and wrote off the sighting as a mistaken identification.

Meanwhile, Col Nguyen Van Than may have thought his operation was discovered, but after several days of inactivity, his effort continued. Within days, replacement supplies began to flow in from the nonoperational SA-2 site at Tam Luong, just across the border in North Vietnam.

> "The sections made up of [radar and fire-control] equipment all stopped at assembly areas and were carefully camouflaged before dawn each morning. In spite of these careful and tight arrangements, the battalion still suffered equipment losses and personnel casualties. An enemy AC-130 detected a section as it passed Se Mu and knocked out

one tracked vehicle and a fire-control computer van. Two TZM missile transport vehicles loaded with missiles were attacked and knocked out by the enemy just before they reached their assembly area, and both drivers were killed." People's Army of Vietnam Air Defense Service, People's Army Publishing House, Hanoi, 1994. English translation by Merle L. Pribbenow CIA Operations Officer (Ret.)

An SA-2 missile and reloader-transporter moving over rough terrain. (photo credit unknown)

The *May Huc* site was declared operational on 24 February 1972. It was an unconventional arrangement of two missiles with only about 50 degrees traverse capability because of the surrounding jungle and hills. The limitation meant the missiles had a relatively small launch window, aimed between two peaks, directly where AC-130s entered the trail network.

On the night of 27 February, Spectre 10, commanded by Capt Greg Seitzler, was firing on their sixth target of the evening in VR-4A as they unknowingly approached the secret *May Huc* site. As soon as the EWO's RHAW receiver indicated radar activity, he activated ECM. Capt Seitzler reacted immediately by rolling out of orbit to exit the area. Meanwhile, the SA-2 missile crew followed their standard three-plus minute engagement sequence. Fortunately, this gave Spectre 10 more time to fly away from the site. A short time later and still within range, the EWO announced an SA-2 lock-on and launch. Right Scanner Sgt John Wood saw two missiles emerge from the jungle and called, "SAM three o'clock!" over the interphone system.

> "I was in the scanner's seat and saw them launch at the three o'clock position. When they were fired, it was like two explosions on the ground. When they reached altitude, it was like a white headlight came on. One disappeared and the other turned towards us. I quickly lost sight and SSgt Larry Bower, our IO had the call. It was quick I am sure, but waiting to die seemed like an eternity."
> **Sgt John Wood**

Spectre 10 estimated the launch site was at least ten miles away; but, at top speed over 1,800 miles per hour, the missiles closed in on the aircraft quickly. Since one missile appeared to be guiding towards the aircraft, Capt Seitzler pushed forward forcefully on his yoke to make a descending turn from 10,300 ft. to 8,500 ft., hoping to break lock. During the maneuver, zero-G conditions caused gunners, ammunition, and some equipment to float in midair. Seconds later, one missile air burst at 10,500 ft. and the other at 6,500 ft. After the missiles passed, Capt Seitzler pulled out of the dive which caused everything that was floating in the aircraft to slam to the floor, injuring two gunners in the process. Though the entire engagement may have seemed like it took an eternity, it was all over in less than a minute. Fearing another launch, Capt Seitzler wisely exited the area.

The report of an SA-2 engagement in VR-4 was again doubted by 7th Air Force Intelligence officers who insisted the crew misidentified what they had seen. There was no doubt whatsoever in the minds of the AC-130 crew engaged by the site. The next day, O-2 and OV-10 FAC pilots searched for the site but found no firm evidence. Lacking proof, AC-130 missions continued to be scheduled into VR-4. Six days later, Spectre 10 experienced a similar event in VR-4, but their reports of an SA-2 engagement continued to be swept aside as an erroneous sighting.

While reports of an SA-2 firing in *Steel Tiger East* VR-4 fell on deaf ears at 7th Air Force, the *Pave Aegis* crew flew nearly every night into *Barrel Roll* engaging every truck they could find with their 105mm gun. The effects of a 105mm HE projectile were spectacular, and it was obvious 40mm BDA criteria was not appropriate for the 105mm. A single direct hit from a 105mm HE projectile left no doubt as to the condition of a truck. Fenders were blown off and large pieces of the vehicle were torn away. A hit on the cab of a truck essentially erased the front of a truck. Wheels and tires sometimes separated from their axles and bounced into the jungle. A hit in the bed of a truck blew supplies off the truck and sometimes left a gaping hole in the bed. Drivers were shredded by fragments or blown to pieces if hit directly. A near miss as far away as 35 feet peppered trucks with hundreds of high-velocity metal fragments. If a fire did not result, 40mm Misch Metal was used to ignite gasoline leaking from perforated tanks. On the night of 4 March 1972, a single 105mm round destroyed three co-located trucks, a capability not seen before, even with 40mm Misch Metal. The 105mm was magic for Spectre crews but the 7th Air Force refused to change BDA kill criteria, clinging to their position that a truck had to burn or explode to receive credit for a kill.

> "...rounds are wasted by continual firing at the target in an attempt to get a direct hit or produce a blower or burner....another factor to consider in that the combat environment does not usually permit prolonged firing orbit around the same target. We must maximize destruction damage in minimum time, then move to avoid accurate AAA reaction."
> Project CHECO, Pave Aegis Weapon System (AC-130E Gunship), HQ PACAF, July, 1973.

The effectiveness of the *Pave Aegis* 105mm cannon did not go unnoticed by PAVN leadership and they were aware the aircraft could fly higher than the effective range of 37mm AAA. To counter their newest adversary, the 591st Regiment established more radar-directed 57mm AAA sites near strategic intersections. Since radar guns did not need to use tracers, the only sign a crew that was being engaged saw were two large circular flashes radial to the bore axis every time they were fired. Sometimes referred to as Mae West flashes, the gun positions were exposed once they began firing, no matter how well they were camouflaged, but crews were unable to see 57mm projectiles coming at them, so they were unable to maneuver to avoid being hit. The trail was even more dangerous near Ban Laboy Ford and Base Area 604 where 85mm and 100mm guns were active.

March 1972 proved to be a very busy month. *Pave Pronto*, *Pave Spectre*, and the sole *Pave Aegis* aircraft were flying missions every night, often redirected from armed reconnaissance missions to TIC support in eastern Cambodia. One of the most effective missions that month was in support of the Cambodian 10th Group Battalion Infantry. When the mission was complete, Chief of Staff Colonel Phin Oum and his fire support officer Capt San Sok (call sign Rustic Sam) visited the 16 SOS to thank them for their support.

Ho Chi Minh Trail defenses became more prolific and effective as the dry season continued. Gunship crews did their best to keep one step ahead, but the C-130 performance envelope was a physical limitation that could not be exceeded. Operational altitude for all aircraft was increased to 10,000 ft. which reduced but did not eliminate the probability of getting hit by 57mm AAA. Such was the case on 4 March 1972 when *Raid-Kills 'Em Dead* took a direct hit from 57mm AAA. That night, Capt David Hobgood's crew (call sign Spectre 17) was tasked into *Steel Tiger East* VR-5. While working their third target of the evening, IO Airman First Class (AIC) Roger Dale scanned for AAA from the open ramp when he saw two large flashes on the ground. Shortly afterward, two bright flashes lit the sky, the aircraft shuddered and the sound of an explosion and hard metal-to-metal impact was heard. Capt Hobgood performed a controllability check while crewmen checked in on the interphone system. Though the aircraft was still flying normally, Capt Hobgood was painfully aware his aircraft was hit badly.

> "The cockpit filled with black smoke from the blast. It was so thick that it was difficult to see the co-pilot or the flight engineer. It seemed to last forever but dissipated in a minute or so. I never changed the airspeed or flap setting on the return trip to Ubon. In fact, I never changed the flap setting at all and only slowed to landing speeds as I approached the runway. Hats off to the crew in the back. They took care of everything."
> **Col, (Ret.) David Hobgood,** pilot.

Nothing, however, could be described as being normal in the sensor booth. A 57mm projectile had exploded on the fuselage directly beneath the booth, blowing a jagged three-foot hole in the floor of the booth. The pressure of the explosion stunned everyone in the booth and knocked the booth door off its hinges, blowing it about fifteen feet, ultimately landing on top of AIC Dale.

> "I saw the muzzle flash directly below us and before we knew it, we were hit. The booth door blew off and landed on me with the heavy flak curtain still attached. I carefully slid out from underneath it and noticed the air was so thick with hydraulic mist I could hardly breathe." **AIC Roger Dale**

Raid Kills 'Em Dead was hit hard. The blast knocked FCO Capt Lee Kun out of his seat and LLLTV operator Capt Gibson was hit in the neck by a shell fragment. IR operator Capt Gary Chandler's seat was knocked off the floor and he fell into the hole. Fortunately, he fell in feet first and his parachute wedged on the jagged metal, keeping him from falling out of the aircraft. Though Capt Chandler's armor-plated seat likely saved his life, he was hurt badly. Stunned, but aware Capt Chandler could slide out of the aircraft at any moment, student LLLTV operator 1st Lt Ed Bunch, EWO 1st Lt Lee Derosa and AG SSgt Ken Schaff rushed to Capt Chandler's aid and pulled him from the hole. When 1st Lt Bunch examined Capt Chandler for wounds, he was in a great deal of pain and his feet were bleeding profusely. While 1st Lt Bunch administered first aid, *Raid Kills "Em Dead* headed towards Ubon.

> "I recall a sudden brilliant flash of light. I was lifted out of my seat by the blast, my seat was blown back and I dropped down into the hole in the floor created by the exploding shell. I was up to my armpits with my parachute wedging me in and preventing me from falling further (and completely out) of the aircraft…I struggled to pull myself from the hole I was in but the parachute that saved me by wedging me in now complicated my extraction. It took three crew members to pull me out: Lee DeRosa, Ed Bunch, and Ken Schaaf." Col (Ret.) Gary Chandler, IR sensor operator

While the crew of Spectre 17 saw heavy damage on the inside of the booth, one of their escorts flew alongside to inspect for signs of a wing fire, but there were none. Since the condition of the landing gear could not be determined, Capt Hobgood ordered the lead gunner, TSgt Thomas to direct the gun crew to jettison the flare launcher, throw out all ammunition and prepare for a crash landing. The flight back to Ubon was uneventful with exception of a constant rush of air coming into the booth. On final approach, Capt Hobgood, copilot 1st Lt Buchout and flight engineer SSgt Norris prepared for a crash landing while AIC Dale helped the rest of the crew strap in on the ramp. Fortunately, *Raid*

Damage from a 57mm hit on aircraft 56-0509 and aircraft 69-6574. (USAF Photos)

Kills 'Em Dead's landing gear and tires were undamaged and they landed uneventfully. When crewmen evacuated the aircraft, Capt Chandler was drifting in and out of consciousness as he was carried to a waiting ambulance. Once at the base hospital, doctors determined he needed emergency surgery to save his feet, but there were no surgical facilities at Ubon and no medevac aircraft available. Concerned Capt Chandler might not ever walk again, 16 SOS commander Col Harry Canham directed a gunship crew to fly Capt Chandler to Da Nang for emergency surgery. It took some time to patch *Raid Kills 'Em Dead* for a one-time flight to Warner Robins Air Force Base, Georgia for major depot-level repairs.

Another surprise occurred on the night of 6 March 1972, when the *Vietnam Peoples Air Force* (VPAF) flew Russian-built MIGs into *Steel Tiger* for the first time. Because an AC-130 would be an easy kill for a MIG, *Moonbeam* immediately instructed all *Spectre* aircraft to clear the area. Unfortunately, it was too late for Spectre 20 who had already detected an active search radar merely fifteen miles away and closing. Without hesitation, Spectre 20's pilot descended to two thousand ft. and headed west while US fighters pursued the MIGs. When the MIGs were chased off, AC-130s returned to their assigned VR sectors, only to be chased out by MIGs again.

> "Suddenly a loud broadcast boomed over our radio: "This is Red Crown on Guard---Bandits, Bandits, Blue Bandits, Squid 240 for

> 40, heading southwest." After a quick glance at his maps, the Navigator called "That places the MiG about 45 miles NW of us and he's coming our way! Recommend we head back towards Udorn". The pilot responded by jamming the throttles forward and standing the aircraft on a wingtip. In short order, we were headed south at 280 knots—100 knots over our configuration-constrained AC-130E's Never-Exceed Airspeed!! Nerves started to fray." **Maj (Ret.) Gerry Stiles**

The possibility of an SA-2 site operating in Laos was still on everyone's mind. Regardless of what HQ 7th Air Force thought, everyone at Ubon was confident there was an SA-2 lying in wait for them west of Tchepone, but missions continued to be sent into the area nevertheless. On 9 March, Spectre 21 was working in VR-4 at 10,500 ft. when another SA-2 launch was detected and witnessed. Spectre 21's EWO activated the ALQ-82 jamming pods, evasive maneuvers were initiated, and chaff was dispensed. During their defensive maneuver, a missile passed over the aircraft, started to descend, then detonated about three thousand ft. above and to the right of the aircraft.

> "Spectre 21 (AC-130) was at 1547/10629 heading 330 degrees at an altitude of 10.5M feet when the APR-25/26 activity and launch lights were activated with receipt of related audio (the only RHAW indications noted).

> The AC-130 began a left turn away from the area of suspected SAM sites when a SAM was noted above the aircraft at approximately three o'clock and appeared to guide down towards the aircraft. Spectre 21 went into a dive, turned on his ALQ-87 pods (jamming against the beacon downlink and TWS frequencies), and dispensed one ammo box of QRC-530 chaff and two canisters of RR-148 chaff." Official declassified 16 SOS records.

Though Spectre 21's F-4E escort (Gunfighter 32) witnessed and confirmed the SA-2 launch, 7th Air Force Intelligence listed the launch as "probable" since, after considerable search and reconnaissance, they could not locate the site. Nevertheless, there were far too many reports for the 7th Air Force to continue to deny a site existed.

By the middle of March, trucks were getting progressively harder to locate, so crews were apt to fire on any legitimate target they could find. Though crews had orders to not engage AAA sites, sometimes they shot AAA guns when no trucks could be located. Such was the case concerning a gun site called "Two-Shot Charlie" which fired only two rounds of inaccurate 37mm whenever a gunship passed by. After each mission, crews laughed when they reported "two rounds inaccurate 37mm AAA" at that location. One night, a crew fired one 105mm round at Two Shot Charlie just after he fired. The next night, as another crew passed the site, they were engaged with a high volume of accurate fire.

Finally, it was realized Two Shot Charlie was signaling other gun sites that an AC-130 was approaching.

HQ 7th Air Force was impressed with the performance of *Pave Aegis* and agreed to allow a second AC-130E to be modified with the 105mm gun. On 15 March, just as the second gun system arrived at Ubon, the *Pave Aegis* aircraft was hit by 57mm AAA. *Iron Maiden* was severely damaged, but the crew managed to safely recover at Ubon. Since the aircraft needed extensive repairs, the 105mm gun system was removed and installed on 69-6571 (*the Destroyer*).

Ten days later, aircraft 0014 received a direct hit between the engines in the right wing. Fortunately, shock isolating foam in the tank prevented a fuel explosion, no fire developed, and the crew safely recovered at Ubon.

When PAVN armor was spotted entering Laos, 40mm Armor Piercing- Tracer (AP-T) rounds were sent to Ubon. While there were doubts of the round's effectiveness against armor because of velocity degradation at range, the round was evaluated in combat. After Spectre 12, fired 596 rounds of AP-T at one tank without success, use of the round was discontinued.

By mid-March, Col Nguyen Van Than remained focused on shooting down an AC-130 at *May Huc*, but he had a dilemma to solve. Because of two unsuccessful engagements, he believed the secret site was compromised and would be destroyed by American bombers. Henceforth, on 20 March, he directed the entire site be relocated to a pre-planned secondary location called *May Huc II*. Of course, this site was also heavily camouflaged to blend in with the jungle. A large crew of workers kept freshly cut plants on the missiles and equipment while the 67th Missile Battalion moved to the *May Huc II* site. During their move, an AC-119K *Stinger* gunship spotted two tracked vehicles then opened fire with 20mm HEI. The attack disabled both vehicles and a radar headed to *May Huc II*, but nobody knew the significance of the vehicles...it was just two more vehicles on the trail. Within a few days, all equipment was transferred and damaged equipment repaired. *May Huc II* became operational on 23 March 1972.

That night, Spectre 06 escorted by F-4s passed through the familiar trail entrance point in VR-4 between Hills 646 and 159. Unknown to them, they were passing directly into range of the *May Huc II* launch site. After Spectre 06 passed the site, radar operators powered on their *Fan Song* radar with hopes of a successful rear-aspect engagement. Fortunately, Spectre 06's EWO detected SAM activity, immediately applied ECM and dispensed chaff. PAVN missile men were hampered by ECM and chaff, so by the time a lock-on was achieved, the gunship was about 11 kilometers away from the site. As soon as two missiles were fired, Spectre 06's pilot maneuvered and successfully broke lock. Ultimately, both missiles passed overhead and detonated some distance from the aircraft. Spectre 06 then exited the area at 3,500 feet altitude to prevent another lock-on. After landing and debrief, 7th Air Force Intelligence finally accepted the fact there was a rogue SA-2 site hidden in the jungle of Laos. In response, an F-105G "*Wild Weasel*" SAM suppression aircraft was assigned to patrol the area in question while AC-130 crews hunted for trucks. The F-105G was specifically equipped to destroy or defeat SA-2s with ECM, AGM-178 Standard ARM, and AGM-45 *Shrike* Anti-Radiation Missiles that homed in on active *Fan Song* radar emissions. Wild Weasels took advantage of the time lag by launching Anti-Radiation missiles as soon as a *Fan Song* radar began emitting. Most believed the addition of the F-105G into *Steel Tiger East* VR-4 would either suppress the rogue missile crew to prevent firing or kill the site a soon as it began to emit. There were limitations to the theory, however; the F-105G had to be within range of the site to be effective, and nobody knew exactly where the site was.

Meanwhile, 275th Missile Regiment Commander Col Nguyen Van Than was disappointed his men had failed to shoot down an AC-130 on three separate occasions. He studied each engagement and concluded the primary reason for failure was a reluctance to engage the AC-130 head-on. Furthermore, the missile crew activated their *Fan Song* radar for an extended period before each launch, which gave the AC-130 crew ample time to employ effective countermeasures.

> "During the course of combat operations, the 275th Regiment had scored victories but had not fully exploited its firepower. It had not yet successfully engaged B-52's and still demonstrated fear of approaching AC-130's. The regiment had engaged AC-130's on three occasions, but on all three occasions,

> the AC-130's had struck back at the regiment. For this reason, a number of cadre and soldiers over-estimated the AC-130 and were afraid to engage AC-130 aircraft." History of the 377th air defense division" Colonel [Thuong Ta] Nghiem Dinh Tich and Colonel Ky People's Army Publishing House, Hanoi, 1998 [Luu Hanh Noi Bo] Page 174. English translation by Merle L. Pribbenow CIA Operations Officer (Ret.)

Col Nguyen Van Than challenged his men to face their fear of the AC-130 and engage it on approach instead of flying away. He deemed the element of surprise critical, so the *Fan Song* radar needed to be kept turned off as long as possible to prevent early warning. Since typical AC-130 altitude and approach heading to the site was known, radar data was pre-prepared to scan that small sector of the sky. In effect, instead of relying on *Fan Song* radar operators to find the AC-130, the aircraft would fly into an electronic ambush where rapid lock-on and launch could be made before the gunship could apply effective countermeasures.

> "The battalion commander decided that he would use Radar 1 to pick up the enemy target and determine AC-130's flight path and flight data. When the target reached effective missile range, Radar 2 would be switched on and would engage quickly using pre-prepared data [provided by Radar 1], selecting

> the timing of the missile launch so that the missile would encounter the AC-130 when it was at the closest point to our launch position." People's Army of Vietnam Air Defense Service, People's Army Publishing House, Hanoi, 1994. English translation by Merle L. Pribbenow CIA Operations Officer (Ret.)

On the night of 29 March 1972, *Prometheus* (call sign Spectre 13), commanded by Maj Irving Ramsower, taxied out of its revetment, marshaled by the aircraft's crew chief, Sgt Tom Combs. After a brief sensor alignment and tweak, they rendezvoused with their fighter escorts and headed for *Steel Tiger East* VR-4. Once in their working area, F-105G (call sign Suntan 02) was already patrolling the area, ready to locate and fire on the SA-2 site as soon as detecting *Fan Song* radar emissions. Spectre 13 was escorted by a flight of F-4s as Maj Ramsower's crew began their mission with a spiral search of Route 9. While sensor operators Capt Halpin and Castillo searched for vehicles, EWO, Maj Howard Stephenson kept his eye for signs of radar activity. SSgt Caniford scanned for threats while hanging halfway over the open ramp. Though the entire crew was aware of a potential SA-2 in VR-4, they pressed on with caution, as they always did. Unfortunately, they had no idea they were headed directly into a SAM trap situated at the *May Huc II* site. On the ground, the SA-2 missile crew heard the familiar sound of an AC-130 approaching them. Every orbit brought Spectre 13 incrementally closer to the site, but as planned, the missile crew intentionally kept their radar off as to not reveal their position. Meanwhile, Suntan 02 was patrolling twenty-eight nautical miles away from Spectre 13, looking for telltale radar signatures.

Only after Spectre 13 flew directly into the narrow, pre-determined SA-2 engagement window did the Missile Guidance Officer direct the *Fan Song* target tracking radar to be switched on. Immediately, Maj Stephenson announced SAM radar activity while applying electronic countermeasures and dispensing chaff, but it was too late. Within seconds, the missile crew received a lock-on, activated missile guidance, and fired. Large flashes of light lit the jungle as two SA-2s streaked off their rails, laying down tracks of fire and smoke as they rose skyward. There was no time for Maj Ramsower to perform effective defensive maneuvers before the aircraft was hit directly by one of the missiles.

> When the target reached a range of 13 kilometers, the missile guidance officer shouted, "Target is in launch range. Switch to combat circuit." All three guidance operators called out that their circuits were good. The missile guidance officer pressed the launch button, and two missiles leapt from their launchers and sped toward the AC-130. Both missiles guided properly. History of the 377th air defense division" Colonel [Thuong Ta] Nghiem Dinh Tich and Colonel Ky People's Army Publishing House, Hanoi, 1998

> (Internal Distribution Only) [Luu Hanh Noi Bo] Page 174. English translation by Merle L. Pribbenow CIA Operations Officer (Ret.)

About twenty-eight miles away, Suntan 02 detected *Fan Song* radar activity the same time Spectre 13 did but was out of range. Spectre 13's F-4 escort, (call sign Barfly 03) observed the thirty-second engagement that resulted in Spectre 13 being hit directly by the second missile fired. The impact flash and subsequent explosion were so violent that it was seen by Spectre 10 working in an adjacent VR sector more than twenty miles away. When hit, Spectre 13 broke apart and spiraled towards the ground, with the main wreckage impacting the ground about 300 meters from route 9 near the village of Nabo. Numerous secondary fires were seen, but no parachutes or radio contact was made with any member of the crew however, survival radio beacons were heard for about 15 minutes.[54]

> "On 29 March 1972, the 67th Missile Battalion shot down one AC-130 that crashed on the spot, thereby fulfilling its assigned mission in an outstanding manner and contributing to our successful defense of Group 559's "Quang Trung" transportation campaign." People's Army Publishing House, Hanoi, 1994, Page 118. English translation

54 The PAVN was known to activate survival radio beacons to lure additional aircraft into firing range but was not known to transmit voice messages.

by Merle L. Pribbenow CIA Operations Officer (Ret.)

Back at Ubon, Sgt Combs waited patiently for his aircraft to return, but after Spectre 13 was overdue for more than an hour, he knew something was wrong. During debrief, Suntan 02 confirmed two SA-2s fired however, Barfly 03 reported three. Nevertheless, *Prometheus* was down and the fate of Spectre 13's crew was in question.

> "I have a vivid memory of returning to my empty revetment along with fellow crew chief Sgt Rhett. The area instantly became large and ghostly. Every stain, every tire mark took on a sudden significance. I remember trying to "wish" my gunship back. My hopes were that somehow all this was just a mistake and my aircraft and crew would be arriving any moment." **SSgt Tom Combs,** crew chief.

The area was considered too dangerous to mount a search-and-rescue effort but was monitored electronically for 24 hours. Unfortunately, there were no beacons or contact with any member of the crew. With no signs of life detected, the entire crew was listed as Missing In Action (MIA) and a massive search effort was mounted to find and destroy the launch site. Meanwhile, at the *May Huc II,* site fearing discovery, Col Nguyen Van Than ordered the entire system shut down, camouflaged, and temporarily abandoned.

The entire squadron was shocked by the loss of Spectre 13, but missions continued to be scheduled. Col Canham was concerned there might be more SA-2 sites in Laos and VR-4 was placed off-limits. The next night, one of about a dozen AC-130s flown into *Steel Tiger East* was *The Destroyer* (call sign Spectre 22), Commanded by Capt Waylon Fulk. As the only *Pave Aegis* aircraft on station, *The Destroyer* entered *Steel Tiger East* VR-7 under a full gunner's moon. After resolving a minor fire-control computer issue, three trucks were located and destroyed. Continuing their search for several more hours, Spectre 22 noticed targets were scarce. Near the end of their mission, sensor operators spotted a truck about three and a half miles north-northwest of their position. Rolling in on the target at an altitude of 10,500 ft., they opened fire just as a 57mm AAA gun began firing at them. Undeterred, *Spectre 22* continued to engage the truck but was unable to score a hit.

> "Around 2145 hours we started taking some 57mm ground fire at our 11 o'clock position at one location in the orbit. On each proceeding orbit, the ground fire was getting progressively closer. We would take between 20 to 30 rounds of 57mm on each orbit. Around the 3rd or 4th orbit we took a volley of 57mm that passed so close to the aircraft that it lit up the tail and the ramp."
> **Sgt Robert Jacobs,** Aerial Gunner.

Spectre 22 continued their attack but were on low fuel. With each orbit, AAA fire was getting progressively closer, but the truck had not yet been hit. Since their replacement aircraft had not arrived, Capt Fulk decided to make one last orbit before returning to base. To throw off AAA gunners' aim, he rolled out of orbit temporarily, then back in, but as soon as he reacquired the target, SSgt Tony Denoi observed tracers fanning their position. Denoi called out, "11 O'CLOCK HOLD WHAT YOU GOT," hoping the rounds would pass without hitting *Spectre 22*. A few seconds later, *The Destroyer* took two direct hits, one on the fuselage and the other just outboard of the #3 engine. Capt Fulk immediately turned towards Ubon while each crew member checked in on the intercom system with their status.

Thankfully, there were no injuries, but the right wing was on fire and sparks were coming from severed electrical wiring from an ALQ-87 ECM pod. Capt Fulk broadcasted a MAYDAY call while burning fuel poured from the wing and flames trailed the aircraft as far back as the horizontal stabilizer. Flight engineer SSgt Cleo Davis shut down the #3 engine, but the fire kept building, fed by leaking fuel.

Capt Fulk wanted to get as close to Ubon as possible before ordering bailout but realized the right wing could fail at any moment, so he advised his crew they were free to abandon the aircraft at their own discretion. In response, the copilot, Capt Eddie Bolling, jumped out of this seat and headed towards the back of the aircraft. As Bolling hurriedly climbed down the flight deck ladder, he slipped and fell into the 20mm bin and became disoriented. When he regained composure, he jumped from the right scanner's hatch directly into 230 knots of airflow (about 80 knots faster than normal jump speed). During this violent bailout, he bounced off the side of the fuselage, through a cloud of burning fuel, and into the darkness. About the same time, gunner Sgt Bob Jacobs jumped off the ramp, but everyone else remained aboard, with hopes of clearing out of the hostile area below before bailing out. Both Bolling and Jacobs experienced violent parachute opening shock, but were otherwise unhurt and under the canopy as they watched their aircraft fly away. Both men landed close to a major trail that was teeming with hostile North Vietnamese. Obviously, this was not an ideal scenario.

> "When I first hit the ground, I was saying to myself "This couldn't be happening to me", but it was. I thought Charlie would capture me right away because I knew they saw the aircraft on fire and me descending in my chute. I landed in what seemed to be, because of the darkness, a level field of Bamboo, Elephant Grass, and brush about six to eight feet high. I didn't move from my chute because I remembered what they taught us in Survival School, and that is if you ever bail out at night you should wait until first light so you can see where you have landed and get your bearings at that time." **Sgt Robert Jacobs**

As Spectre 22 headed towards Ubon, Capt Fulk and SSgt Davis tried to extinguish the wing fire while the rest of the crew stood on the ramp, ready to bail out. Capt Fulk tried everything he could to extinguish the wing fire, including a diving maneuver to blow it out, but the fire persisted. During this time, a B-57 pilot escorting Spectre 22 noticed the fire had burned a hole through the extended right-wing flap and the center portion of the right wing was on fire. Knowing the wing could fail at any moment, Capt Fulk activated alarm bells and ordered his crew to bail out. Crewmen began jumping from the ramp in pairs spaced about five seconds apart. Once certain all other crewmen had jumped, Capt Fulk engaged the autopilot, jumped out of his seat, and ran to the back of the aircraft. He and SSgt Denoi were the last to bail out. Shortly afterward, the aircraft entered a descending turn, and crashed into the jungle near Saravane, Laos, creating a large fireball.

Spectre 04, working in the next sector, heard the mayday call, pulled off target, and headed for the crash site. They arrived to find crewmen dispersed in several groups across heavy jungle terrain and the first two to jump separated from the main group by about 30 miles. Thirteen survivors were in *Steel Tiger West* in a relatively safe area about 75 miles from Ubon but the other two were right in the middle of a major PAVN controlled area of *Steel Tiger East*. Spectre 04 began contacting downed crewmen one-by-one via survival radio to locate their positions to pass to rescue forces. The 40th ARRS and *Air America* were alerted for an immediate rescue mission; however, neither unit was night capable, so a rescue operation beginning at sunrise was planned. Remarkably, the entire crew survived bailout but several were injured. Since a night-time rescue was not possible, crewmen had no choice other than to tough it out until daylight.

At 0455, six HH-53 *Jolly Green* helicopters from the 40 ARRS and 37th ARRS took off, heading for their initial orbit point. Once there, they joined A-1'S of the 56 SOW and F-4s from the 388th and 8th TFW in what was already being called the "Great Spectre Easter Egg Hunt." The task of rescuing fifteen crewmen spread out over 30 miles was not an easy task. An HC-130 crew (call sign King 27) commanded by Capt David Preston, controlled the rescue operation. The overall plan was based on rescuing the main group first, then the two located in *Steel Tiger East*. The key to success was the rapid location of the dispersed crewmen and control of the overall rescue force. At first light, Raven 40 located and rescued Capt Fulk and SSgt Denoi. At 0725, Jolly 53, commanded by Maj Kenneth Ernest, flew to the survivors. Pararescue men SSgts Charles Morrow and Gary Osborne were lowered to the crewmen to tend to their injuries and assist in hoisting them onboard. Once onboard, Jolly 53 returned to Ubon. In sequence, Jolly 62 picked up one survivor, Jolly 63 picked up two and Jolly 60 picked up eight.[55] With 13 rescued and headed for Ubon, an

55 Jolly Green pilots during this mission included Capt Dale Stovall, 1st Lt Steve Connelly, and Capt Bennie Orell. All three served in special operations for many years. Stovall retired as a Brigadier General while Orell and Connelly retired as Colonels.

Air America UH-34D helicopter commanded by Mr. Bruce Jachens and an H-45 commanded by Mr. Allen Cates were called in to pick up Capt Bolling and Sgt Jacobs.

> "Arriving in the zone, contact was made with both crewmembers. No one hesitated. They needed to get in and out as soon as possible. Bruce held his altitude and acted as SAR for H-45 who picked up "B" uneventfully. As soon as H-45 was safely up and away, Bruce descended to pick up "M" while H-45 acted as his SAR." **Mr. Allen Cates**

The *Air America* crews recovered both crewmen without drawing enemy fire. Their courage is especially commendable considering they were flying unarmed, antiquated helicopters that could have been easily shot down. Survivors were flown to Pakse for medical evaluation, where they remained until the following day when an HH-53 crew flew them back to Ubon.

When rescue helicopters arrived at Ubon, crewmen were met by an entourage of fellow *Spectres* champagne bottles in hand. Though everyone survived, three crewmen suffered serious injuries. SSgt Tony Denoi suffered a broken neck, gunner Sgt Dave Tisron incurred a broken leg, and the copilot, Capt Eddie Bolling suffered injuries as a result of the impact with the aircraft upon bailout from the right scanner's escape hatch. With the loss of two AC-130s and one entire crew in two days, morale was somber, to say the least, but missions continued to be scheduled.

Sgt Robert Jacobs greeted by squadron members as he stepped off a rescue helicopter at Ubon. (USAF photo)

Spectre 22 Crew			
AC Capt Waylon Fulk	CP Capt Eddie Bolling	TN Capt Richard Cardosi	LLLTV Lt Col Omer Lentz
BC Capt Clarence Johnson	FCO Maj Dallas Dixon	FE SSgt S.Sgt. Cleo Davis	IO SSgt Tony Denoi
AG SSgt Mike Frost	AG Sgt. Dave Tisron	AG Sgt Robert Jacobs	AG Sgt Mike Tesch
IR Capt Robert Rayburn	AG MSgt Jim Thrasher	AG Sgt Stan Reeves	

Group 559 considered the downing of two AC-130s back-to-back a great victory. Truck drivers who had been picked apart by AC-130 crews for nearly five years were especially relieved.

> "This was a very significant victory.... the entire Trail network was happy, but happiest of all were the truck drivers. Just one hour when AC-130's did not operate over our chokepoints was both precious and rare." History of the Truong Son Forces, English translation by Merle L. Pribbenow CIA Operations Officer (Ret.)

Meanwhile, HQ PACAF was deeply concerned the threat level in Laos had become too high to continue to hunt trucks over the Ho Chi Minh Trail. Though not fully understood at the time, *Prometheus* and *The Destroyer* were shot down two days prior to the beginning of a major PAVN offensive to seize control of Vietnamese cities including Hue, An Loc, Kon Tum, and Quang Tri. This multipart offensive was later named "The Easter Offensive" by US Forces. By the evening of 30 March, forty thousand PAVN troops and six hundred armored vehicles were pouring into South Vietnam via Laos and Cambodia. The 7th Air Force mounted a high-level effort to halt the PAVN incursion, but the scale of the operation was overwhelming. Every attack, fighter, and bomber aircraft available flew missions against the mass infantry advance. At the same time, gunship crews flew TIC missions in support of friendly forces defending the major population centers under siege. Meanwhile, PAVN troops and supplies continually poured down the secret "K-Trail" undetected, under the cover of darkness.

The arrival of April 1972 and the onset of the rainy season that normally resulted in scaled-back AC-130 operations did not occur. Since the Easter Offensive was in full swing, AC-130 missions continued into the rainy season, regardless of unfavorable weather conditions. The 16 SOS was assigned the task of supporting US and ARVN forces day and night including Saigon Capital Defense in response to increased *Viet Cong* attacks in and around Saigon. These missions were commonly known as Saigon Turns. Many times, AC-130 crews landing at Tan Son Nhut were fired upon with small arms on final approach or caught on the airfield during mortar attacks, but no significant damage occurred.

> "The infamous "Saigon Turn" went on for a long time. We would take off after dark from Ubon and fly down to Saigon. Flying airborne alert for around 4.5 hours and then land at Tan Son Nhut. Reload ammo, fuel, and get some breakfast at the chow hall. Take off and fly another 4.5 hours on alert and land back at Ubon around dawn." SMSgt (Ret.) Frank Hanna, gunner.

> "...we were keeping an aircraft down at An Loc just about twenty-four hours a day. Also, the Kontum area was quite heavy and also in the Saigon area, but our primary concern was An Loc, Kontum, and Quang Tri. We were more or less divided between those three areas on just about a twenty-four hour basis." Project Corona Oral History of Lt Col Loyd J. King, 1973.

During this period, the hunt for the missile site that shot down Spectre 13 was thorough but unrevealing. For about a month, reconnaissance and FAC aircraft searched for the May Huc II site, without finding it. In time, the 67th Missile Battalion disassembled and removed all equipment from the site and crossed the border back into North Vietnam.

The shift from armed reconnaissance on the Ho Chi Minh Trail to TIC in Vietnam was a dramatic tactical change for crews. Crews had never worked within crowded urban population centers and most had rarely worked with ground commanders in the past. The situation was much the same for ground commanders who rarely, if ever, had worked with AC-130s. Another significant change was the shift from night missions in Laos to day missions in Vietnam. Fortunately, the presence of AAA in South Vietnam was much less prevalent than on the Ho Chi Minh Trail, so crews were able to operate between 6,500 and 7,500 ft.

Missions began over An Loc first. To further bolster gunship target acquisition and situation awareness, the IO and Right Scanner were issued binoculars similar to what FAC pilots used. The IO had the best field of view from the open cargo door, so the IO essentially became a visual sensor. From his vantage point, the IO would talk IR and LLLTV sensor operators onto potential enemy targets for a closer look.

> "The first couple of missions that we went in; we would pick up targets on the ground with binoculars—targets that were moving vehicles and troops who were running around in various areas." USAF Corona Project, Oral History of MSgt Arthur Humphrey, 1975.

During these operations, AC-130s were controlled by FACs and ground controllers with Tactical Electromagnetic Impulse Generator (TEMIG) ground beacons. Initially, the drill was always the same; AC-130s arrive overhead to make contact with ground controllers only to be put in holding since most ground controllers were unfamiliar with AC-130 capabilities. Sometimes crews spent their entire mission orbiting at a hold point without being assigned a single target. Crews were frustrated, but during one mission to An Loc, a crew coaxed a ground controller into letting them fire. The ground controllers were impressed with the accuracy and loiter time of the AC-130. From that point on, the good word spread, and soon, AC-130s became the primary TIC aircraft requested.

At least 10 to 12 sorties were scheduled every twenty-four hours. Additional crews were trained to fly the *Pave Aegis* modified aircraft, which was now 69-6573 (*Heavy Metal*). Crews and maintainers worked at maximum effort continuously, day and night. There were no days off and their only downtime was to sleep in preparation for their next mission. Sometimes crews spent hours on alert during pouring rain waiting for the first opportunity to take off. Once airborne, they flew through weather for their entire mission, never seeing the ground. During rare occasions when the weather cleared, crews located and fired on advancing PAVN infantry, vehicles, and tanks in support of US and ARVN forces.

Activity was heavy as PAVN armor, vehicles and troops advanced. Crews knew 40mm HEI and AP-T were ineffective against T-54 tanks, but a direct hit by a 105mm high explosive (HE) projectile proved effective. Since only one *Pave Aegis* aircraft was available, it became the primary tank hunter. During the following days, *Heavy Metal* played a significant role during a mission in Laos while supporting the Royal Laotian Army. On 15 April 1972, five T-34 tanks were destroyed with ten rounds of 105mm HE, halting an attack on Laotian soldiers. On 23 April, Capt

Russell "Rusty" Olson flew the *Pave Aegis* aircraft to Dak To, Vietnam in response to TIC tasking. When they arrived overhead, they were engaged by 23mm and 12.7mm fire that fell short of their altitude. Though the weather was spotty, the crew remained overhead firing on advancing T-54 tanks until low on fuel. After refueling at Pleiku, they returned to continue their efforts. By the end of the mission, they had destroyed seven T-54 tanks before returning to Ubon in the early morning hours.

> "All seven tanks were destroyed with 105mm rounds. The last tank was hit seven times with white phosphorus rounds until it cooked off." SMSgt (Ret.) Ira Roseberry, Gunner.

On 5 June, Royal Laotian Army Brigadier General Soutchai Vongsavanh was so grateful *Pave Aegis* saved him and his troops from certain death that he visited the 16 SOS to present Col Canham and Col Nichols a hatch from one of the tanks blown off by a *Pave Aegis* crew.

AC-130E capabilities continued to be in high demand during the *Easter Offensive*. Ground commanders are known to ask all *Spectre* aircraft arriving on station the same question: "do you have the big gun?"

> "One commander, call sign "Tunnel 10A" quickly exploited the 105mm capabilities. When told a Pave Aegis was arriving, he announced "Great! I've been waiting 24 hours for you to get here and hit this tar-

> get. O.K. all you other guys move off, Big Bertha is here." Project CHECO, Pave Aegis Weapon System (AC-130E Gunship), HQ PACAF, July 1973.

During an attack near Dak Pek on 10 June, a crew worked with a ground commander to drive a group of attackers into a mined killing zone. Once the friendly ground positions were located, firing began. The ground commander corrected fire as close as 500 meters of friendly positions and after firing 62 rounds of 105mm HE and 216 rounds of 40mm HEI, the enemy attack was broken. Later, Ground commanders at An Loc praised the *Pave Aegis* crew and credited them for fifty KBA. It was obvious to all the addition of a 105mm gun was the next incremental improvement of the AC-130 fleet.

> "Communist soldiers fled their bunkers when AC-130 (Pave Aegis) gunships fired on their positions…The American attack plane was the single most effective weapon used at An Loc." **Capt Harold Moffett,** US Army.

> "There was nothing the NVA could do when the Pave Aegis was overhead except crawl into a hole and hope it didn't hit them."
> **Maj. K. A. Ingram,** US Army

MSgt Jake Mercer and Sgt Lonnie Blevins operating the Pave Aegis gun in March 1972. (USAF Photo)

While *Spectre's* capabilities were improving, so were PAVN defenses against it. The PAVN was now equipped with the Soviet-built SA-7 Strela missile. This weapon was a lightweight IR-guided missile designed to be carried and launched by a single soldier. It was capable of precision attack on aircraft flying at altitudes up to 10,000 ft. Unlike AAA or SA-2 sites, there was no way to identify an SA-7 launch area; it was small, passive, and could be anywhere at any time.

The first sightings of SA-7 missiles were surprising since they did not behave like anything crews had seen before. In all cases, the indication of a launch was a small flash on the ground, followed by a blue-white colored corkscrew flight trajectory heading towards the aircraft. Initially, crews used the same defensive maneuvers as they did for AAA, but they quickly learned the missile kept tracking the aircraft as they maneuvered. There were multiple launches and some close calls, but no hits. At the time, the general consensus amongst crews was the SA-7s small warhead was no more effective than 37mm AAA and the SA-7 was more of a nuisance than a credible threat.

On 12 May, a number of crews were flying daylight missions in support of US and ARVN forces. Because of a heavy undercast and inability to see the ground, all AC-130s working An Loc were directed to Return To Base (RTB). Unfortunately, *Heavy Metal* (call sign Spectre 14), commanded Lt Col Richard Kelsay, did not receive the order. As their mission progressed, sensor operators searched for holes in the cloud cover with limited success. As Lt Col Kelsay maneuvered through a broken cloud deck, he noticed AAA tracers streaking through the clouds in front of the aircraft. While Spectre 14's IO, Sgt Ken Felty hung his body over the ramp to see the tracers, he was surprised to see five SA-7 missiles launched in rapid sequence coming right at them. Sgt Felty alerted the crew of inbound missiles at about the same time the LLLTV sensor operator began to track one heading towards the aircraft. Coincidentally, the 2KW light was powered up (in the covert mode), emitting a heat signature.

In a time span of about ten seconds, four missiles streaked past the aircraft, but the fifth locked onto the 2KW light, impacted fuselage, and exploded about ten feet aft of the gimbal. Spectre 14 and Sgt Felty were hurt badly. The warhead blew a jagged hole about two feet across on the left side of the fuselage, projecting fragments through the fuselage and out the right side. Felty was hit in his right arm and hip and more than forty fragments hit his helmet. Three fragments were embedded in his skull and one hit his right eye.

Damage to the aircraft was extreme. Two of the four spars that support the aircraft's tail were completely severed. When crewmen checked to report status and damage to the aircraft, Lt Col Kelsay determined the aircraft was flyable, so bailout wasn't needed. The gunners reported extensive damage to the tail of the aircraft. Noting this, Lt Col Kelsay realized any abrupt flight control inputs could cause the tail to break off.

To prevent structural failure, he reduced airspeed and carefully nursed the aircraft towards the nearest airfield for an emergency landing. Gunners attended to the wounded IO as they watched the tail of the aircraft flex and distort with every flight control input. It took them about an hour to reach Tan Son Nhut. Nearing the airfield, Spectre 14 declared an in-flight emergency and landed without further incident. After shut down, the crew noticed the entire tail section was wrinkled and appeared to be drooping.[56] Crew

56 It took about three months to repair the aircraft.

chiefs inspected the aircraft and were amazed the tail didn't break off…it was a close call.

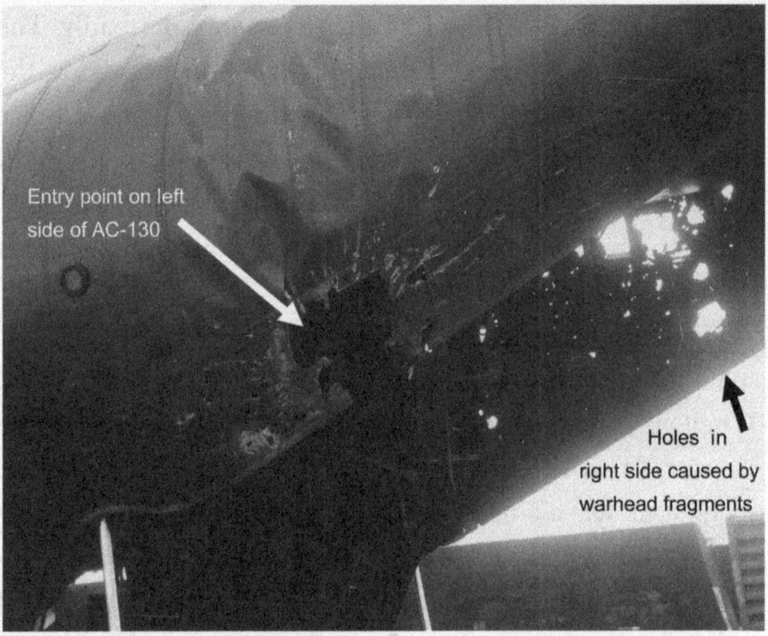

Results of the SA-7 hit on 69-6573 *Heavy Metal* on 12 May 1972. (USAF Photo)

After Spectre 14's tragic episode, it was evident the SA-7 was in a league of its own and much more dangerous than a stray AAA hit. Gunship pilots discovered an abrupt turn into the missile's flight path was the most expedient tactic to defend against the SA-7. During this maneuver, the aircraft's wing shielded hot engine exhaust, causing the missile to lose tracking of the heat source. The first countermeasure adopted was to increase operational altitude to 8,500-9,500 ft. in an attempt to out-range the missile. The second step was to equip the right scanner and IO with M8 flare pistols which they could fire from the aircraft to decoy the missile seeker. The third step was an ad-hoc field modification of LUU-2 illumination flares to be fired from the LAU-74 to ignite immediately upon launch to decoy the missile seeker. The combination of altitude and flares increased survivability incrementally, but a more effective solution needed to be developed and adopted.[57] From that point on, SA-7s were fired at AC-130s during nearly all TIC missions. At the same time, the Gunship Program Office searched for a more permanent solution. Eventually, it was realized an SA-7 would self-destruct after fifteen seconds of flight time, so an increase in operational altitude mitigated the risk of getting hit unless the launch position was almost directly under the aircraft.

Once the enemy realized AC-130 crews had outsmarted their SA-7 tactics by altitude increase and decoy flares, they sent PAVN infantry equipped with SA-7s to a valley near Hue surrounded by mountains that rose to 3,500 ft. This site was specifically chosen to shorten the missiles' engagement time and range to enable a hit on an AC-130. By early June 1972, the platoon was in place near the top of a mountain, capable of a clear shot of any AC-130 that came through the valley. The squad posted a lookout waiting to ambush the first AC-130 that came within range. [58]

---

57 The flare pistols worked fairly well, but when a flare was accidentally launched inside the aircraft and burned through the floor, a better solution was desired.

58 Bare Feet, Iron Will, Zumwalt, P-274

Since there were many missions canceled during heavy rains, the short periods of good weather were taken advantage of to the fullest extent possible. On the early evening of 18 June 1972, *Odin* was within an hour or so of completing a mission when a fresh crew arrived at the squadron, ready to hot turn the aircraft for a second sortie. Because of scheduling and maintenance issues a day prior, the crew was not a typical hard crew, but a mix of crewmen of various levels of experience. Capt Robert Wilson was assigned as student Aircraft Commander under the tutelage of experienced combat veteran Capt Paul Gilbert who was in the right seat. The sensor operators, FCO and EWO had not flown with each other before. Some of the gun crew had flown together and IO TSgt Donald Klinke had completed more than 100 combat missions, but after a minor write-up on a previous mission was getting a requalification flight by IO Instructor SSgt Bill Patterson. Since there were no other gunners available, MSgt Jake Mercer, the highest-ranking gunner at Ubon volunteered for the mission.

After *Odin* landed, refueled, and reloaded, the crew boarded the aircraft. Under the call sign Spectre 11, the crew took off to complete an armed reconnaissance mission in an area near the Laos/Vietnam border called "Oscar Eight." Their mission was to interdict vehicle traffic on the intersection of Route 92 and Route 919, a known strategic choke point near the northern edge of the *A Shau Valley*. This was an area of rugged terrain AC-130 crews routinely hunted because of heavy truck traffic and usually light AAA defenses, but was also within about three miles from a well-protected PAVN support and logistics center.

Spectre 11's objective was to ambush trucks on the intersection and block the strategic intersection for as long as possible without straying into AAA range of guns at the logistics center. Arriving over the target area at 11,500 ft. altitude escorted by fighter escorts under the light of a half-moon, sensor operators began to search. Weather conditions remained surprisingly good while Spectre 11's IOs scanned for AAA from the open cargo door. Within minutes of establishing an orbit, sensor operators located two trucks and the crew prepared to fire, but accurate AAA caused them to break out of orbit. After re-acquiring the targets, SSgt Patterson noticed the telltale ground flash of an SA-7 launch, then announced "STRELA! BREAK RIGHT!" but instead of a hard break maneuver, the pilot initiated a slow roll to the right. Patterson continued to call "BREAK RIGHT, BREAK RIGHT", but the missile kept coming. When the missile was about two or three seconds away, Patterson fired an LUU-2 flare to decoy the missile, but it continued to speed towards the aircraft. In a flash, the missile hit *Odin's* #3 engine, producing a bright flash and muffled explosion. The impact caused the aircraft to shudder, rise slightly, then settle back down. Pilots retained control of the aircraft, and it seemed as if the situation was under control. About ten seconds later, SSgt Patterson scanned the right wing and saw burning fuel and debris trailing from the aircraft. He immediately advised the pilot, "we are hit bad and on fire." Capt

Wilson activated alarm bells directing the crew to prepare for bailout.

> "During this sequence of events, the AC seemed to have had control of the aircraft. I then observed two crewmembers, whom I believe to be gunners, leave the booth area and run out the separating door (between the booth & aft of the aircraft). I then heard the emergency alarm bell ring and the red light came on, indicating that evacuation procedures were in effect." **2nd Lt Robert Reid.**

Less than 30 seconds after being hit, both IOs disconnected their restraining cables and began to clip on their chest pack parachutes. While gunners and sensor operators were assembling behind the booth, the aircraft began to roll hard to the left. Capt Gilbert and Capt Wilson fought hard to keep the aircraft in control but were unable to. Flight engineer TSgt Richard Cole advised the crew they were quickly losing altitude. Sensing the aircraft was out of control, SSgt Patterson looked towards the front of the cargo compartment to see a gunner standing by the 40mm guns with his back on fire while other crewmen were reaching for their parachutes. Suddenly, a large explosion caused the #3 engine to break off. The aircraft rolled hard to the left, then inverted. Now while floating in a negative G condition, SSgt Patterson pushed himself away from the edge of the ramp, scraping his back along the tail section on exit. As soon as he cleared the aircraft, a large explosion caused the right wing to break off. The aircraft tumbled wildly towards the ground while crewmen were pinned against the inner side of the fuselage, unable to move.

In the booth, as soon as 2nd Lt Reid stood up, he was forced into the FCO seat by strong G-forces, unable to move as he watched the altitude indicator and altimeter gauges spin. Certain he was looking death in the eye; all he could do was wait for impact. *Odin's* tumbling trajectory caused extreme wind loads on the tail of the aircraft, causing it to break off. As the aircraft continued to spin and tumble, 2nd Lt Reid and Capt Gordon Bocher were ejected from the aircraft into the darkness.

> "Both the TV and myself did not get very far when the wing fell off. I surmised later we were pinned against the 40mm ammo rack. The G forces were so great that I could not move. I saw bodies all over the aft cargo compartment. Everything was orange. Then the bird exploded and all I saw was a white flash and then black. I reasoned, not felt, that I was falling and I pulled my parachute (backpack) ripcord." **Capt Gordon Bocher**

Patterson, Reid, and Bocher were the only crewmen to escape *Odin* before three major sections of the aircraft crashed onto the jungle floor. While the survivors descended under parachute, Reid's parachute "streamered" for most of his fall as he frantically used a leg movement technique taught in survival school called "bicycling." Fortunately, his chute was fully inflated just prior to landing. Now with all three survivors on the ground, Reid was dangerously close to a manned AAA site as all three survivors concealed themselves in jungle foliage. Though the PAVN controlled Vietnamese jungle was not a welcome place for *Spectre* crewmen, they had no choice other than to sit tight and wait for daylight. During the night, Bocher and Reid made voice contact with one of the F-4 escorts, but Patterson's radio was damaged on bailout. Later, a Nail FAC arrived overhead to pinpoint all three survivor locations. Tragically, there were no signs of any other parachutes.[59]

At first light, a 37th ARRS rescue force supported by A-1 Skyraiders was launched from Da Nang, Vietnam. While the HH-53 "low bird" commanded by Capt Donald Sutton picked up survivors, A-1s dropped BLU-52/B CS Riot Control powder on the periphery of the area to reduce enemy movement. Navy fighters also bombed a nearby airfield to squelch any activity that may be emerging from there. By mid-morning, all three were on board an HH-53, then flown to Da Nang for medical checks.

About three days later, a MEDEVAC aircraft returned them to Ubon where they were met by fellow Spectres, with bottles of Champagne. All three survivors were bruised and beaten by the ordeal, but 2nd Lt Reid suffered a shrapnel wound to his left foot. He received medical attention at Da Nang then was air evacuated to Clark AFB, in the Philippines

59 SMSgt (Ret.) Patterson recollects he may have seen a fourth parachute, but there were no signs of life during subsequent SAR efforts.

for thirty days before returning to Ubon and being placed back on flying status.[60]

Post mission analysis revealed just how fast the sequence of events occurred. Though none of the three survivors were prepared to bail out when the tail broke off, the fact the sensor operators were equipped with backpack parachutes likely saved their lives.[61] Likewise, SSgt Patterson was lucky to survive since he bailed out with only one riser attached to his chest parachute, and the chute was clipped on backward. Fortunately, Patterson was an experienced skydiver and was able to find his D-ring quick enough to deploy his chute.

Determining the fate of the rest of the crew was challenging. Throughout the rescue effort, no radio contact or signs of life were observed. Though the crash site itself was in a relatively unpopulated area, 7th Air Force Intel estimated there were between five to eight thousand enemy troops in the vicinity of the crash site. Concerned a ground search would likely result in heavy casualties, it was ruled out in favor of an aerial search. After six days and no contact or signs of life, twelve crewmen of Spectre 11 were listed as MIA.

During the months of July, August, and September, sixteen SA-7 missiles were fired at AC-130s, but none were hit. Communicating with the Direct Air Support Center (DASC) at Hue, Vietnam was difficult. To remedy the problem, Maj Roy Tingle and Capt Don Ruth were sent to brief and train

60 E-mail from Col (Ret.) Robert Reid

61 It is possible some of the gunners who had not yet clipped on their chest pack parachutes were ejected from the aircraft also.

DASC personnel to resolve as many misunderstandings as possible. After clearing up several issues, they returned to Ubon. From October to the end of December, the 16 SOS flew 551 TIC missions, breaking up numerous ground assaults and yielding 162 KBA. During this time, eight SA-7 missiles were fired at AC-130s, but none were hit.

By the beginning of October, the trails dried out in Laos, enabling full-scale armed interdiction missions to resume on the Ho Chi Minh Trail. Gunship crews returned to familiar territory to find trail defenses much heavier than any time in the past. Crews spent the majority of their mission time dodging 37mm, 57mm, 85mm, 100mm AAA, and SA-7 missiles. Since fighter activity against PAVN forces was heavy, only about half of all scheduled AC-130 missions were planned with fighter escorts. As compared to previous dry seasons, few trucks were found on the trail. The return of hot, dry weather adversely affected *Pave Aegis* which made nearly every takeoff a maximum effort event. Fortunately, the gunship program office had begun an AC-130E improvement program at Warner Robins Air Force Base, GA, called the Combat Required Operating Capability (CROC) program. The effort focused on increasing survivability and effectiveness including more powerful T56A-15 engines, external fuel tanks, additional ceramic armor plating, a GLINT, ALE-20 IR decoy flare launcher system, wing-mounted SUU-42 chaff/flare dispensers, digital fire-control system, and trainable 40mm and 105mm guns. Fuselage modifications included Plexiglas bubbles to protect the right scanner and IO from wind blast and increase aircraft performance during takeoff. The AC-130E/H CROC configuration was envisioned as the ultimate AC-130. The first CROC aircraft (69-6575) was scheduled to be delivered to Ubon in February of 1973.

On 2 November, Spectre 04, destroyed three trucks and damaged seven more in *Steel Tiger East*, but trucks were getting hard to find on the trail. Spectre crews more often found themselves committed to the Saigon defense mission which ran day and night. On 14 November, the first 105mm trainable gun mission was flown by Spectre 07, commanded by Capt Thomas. The hydraulically trainable mount was integrated into the fire-control computer on 69-6570 to reduce the overall gun error budget. Crews found the new mount not only enabled better hit probabilities but eased the requirement for a pilot to fly perfect geometry to achieve hits.

Truck kills increased slightly in December 1972. Spectre 16, commanded by Capt Thompson, was sent to *Barrel Roll* to search for a suspected truck park. Once overhead, they found two groups of trucks, destroyed one group of three, but lost the other group. Searching to the south, they discovered another group of six, ultimately destroying four and damaging one. The biggest mission of December occurred on the 16th when *Bad Company* (call sign Spectre 20), commanded by Maj Davidson, entered *Steel Tiger East* VR 3 and 4. Once overhead, the crew fired on unidentifiable targets which resulted in secondary fires. Suspecting they discovered a truck park or fueling point, Spectre 20 called in their fighter escorts to bomb the area. The jungle lit up with fire and secondary explosions covering an area of about one-third kilometer square. Spectre 20 remained overhead, FAC'ing in two more flights of fighters onto the target before going home.

On 18 December, Operation Linebacker II began, which diverted fighter escorts away from AC-130 support to bombing missions in North Vietnam. Spectre was on its own, hunting trucks in *Steel Tiger* without escorts. Nevertheless, there were three significant missions flown into *Steel Tiger East* VR 5 and 6 on the night of 18 December resulting in twenty-six trucks destroyed, two damaged and several 37mm sites destroyed or silenced.

On 20 December, *Iron Maiden* commanded by Capt Bernie Frakes (call sign Spectre 28) took off for a routine Armed Reconnaissance mission in *Steel Tiger East.* After aligning sensors and tweaking guns, Spectre 28 proceeded towards their assigned VR sector. While searching for trucks, the IO called two rounds of 37MM in the distance. Since the IO had a good mark on the gun site, and they were flying at 10,000 ft. he suggested Capt Frakes go to the site to see if there were any trucks there. Frakes agreed, then headed for the gun site. Just as they arrived, three more gun sites opened fire from opposing directions, trapping them in between. Stuck in a AAA trap, but high enough that 37mm was inaccurate, Frakes decided to engage the gun sites with 105mm HE.

> "Time seemed to slow down as we continued
> to fire the rounds into the gun sites. The AAA

> guns seemed to shut down after we placed a 105MM round near them, so we continued in our fight. The IR then relayed that he was picking up trucks parked around the area. We then started directing some of our fire towards the trucks. The 105MM was driving nails by this time, so we then alternated between the gun sites that were still active and started picking off the trucks. I can recall vividly watching through the HUD as the muzzle flash of a 37MM gun goes off and then a 105MM round hitting it dead-on. No more firing from that site. After several minutes there were no more AAA guns firing and no more trucks. We then decided to clear the area." **Lt Col (Ret.) Bernie Frakes,** pilot.

On 21 December, *Thor* (call sign Spectre 17), commanded by Capt Harry Lagerwall, took off from Ubon for a routine night armed reconnaissance mission in *Steel Tiger East* VR 5 and 6.[62] Weather was clear with 7 miles visibility, and soon after arriving in their working area at 11,000 ft., they quickly destroyed two trucks with 40mm fire. Sensor operators 2nd Lt George McDonald and 1st Lt Delma Dickens were tracking a third truck while Maj Paul Meder, a student IR operator, observed crew procedures and tactics.[63]

62 Maj Francis Walsh Jr., an F-4 pilot from the 497 TFW was onboard for an orientation ride.

63 Maj Paul Meder was a student IR on his first orientation flight.

Just before Lagerwall opened fire on the third truck, a single 37mm gun from inside their orbit began engaging them with accurate fire. Capt Lagerwall broke away from the gun site and flew out of range. Since fighter escorts were not available to bomb the site, Lagerwall remained committed to killing the third truck, so he decided to descend to 7,500 ft. to shoot the gun with the 20mm guns. When Spectre 17 returned overhead at 7,500 ft., Dickens located and tracked the gun site, while Lagerwall established a firing orbit. Just about the time Lagerwall was ready to fire, two more 37mm gun sites engaged the aircraft. During an intense hail of accurate fire, Sgt Stevens called a break to avoid getting hit, but it was too late, Spectre 17 was hit hard on the left wing and fuselage.[64] Capt Lagerwall immediately turned towards Ubon and directed all crew members to check in with their status over the interphone, but the interphone system was intermittent. Though the aircraft was still flyable, it was heavily damaged. The utility hydraulic system and liquid oxygen system were knocked out and both systems depleted. Hydraulic fluid and fuel poured from the wing box onto the floor where it flowed towards the back, collecting at the ramp hinge. Though the smell of fuel permeated the fuselage, not everyone recognized the extent of damage or potential consequences. Flight engineer TSgt James Fuller advised the crew the aircraft was losing fuel at an alarming rate as he attempted to isolate fuel flow from the left wing. By this time, five crewmen had assembled on the ramp while

64 Evaluator IO TSgt John Willingham was administering a check ride to IO Sgt Carl Stevens.

all others remained in their crew stations. Meanwhile, *The Hussy* (call sign Spectre 07), commanded by Capt Bertram McMillen III, was nearby monitoring the situation over the radio when Spectre 17 requested an escort back to Ubon. Ten minutes after Spectre 17 was hit, an explosion inside the left wing caused it to break off, simultaneously igniting fuel fumes that filled the fuselage. A wall of flames quickly enveloped the cargo compartment. Spectre 17 rolled hard to the left and began to tumble out of control. TSgt Richard "Willie" Williams and Sgt Carl Stevens, who were both on the edge of the ramp, managed to bail out.

> "At the time, I was located on the extreme cargo ramp as close to the edge as possible. I managed to kick myself out with some difficulty and before my chute even opened up, the whole sky lit up like there was another explosion, but my back was turned and I did not see it. I had a good chute and after looking around did not see any other parachutes. On my way down, I heard a muffled explosion which I thought was the aircraft hitting the ground. I then noticed three fires on the ground. Extract from official statement of TSgt Richard Williams, 26 Dec, 1972, (declassified on 31 Dec, 1980).

While Spectre 07 continued towards Spectre 17, they observed a streak of light falling from the sky which they thought was a flare being launched. As they arrived over the crash site, they realized the streak was Spectre 17 breaking up in flight and crashing in the jungle. The 40th Aerospace Rescue and Recovery Squadron (ARRS) at Nakhon Phanom Thailand was alerted. A force of two HH-53s, three A-7 *Sandys*, and an HC-130 *King* was launched, but the HC-130 developed an engine problem and had to return to base, so Spectre 07 assumed the role of SAR on-scene commander. Spectre 07 sensor operators Capt Robert Wasson and 1st Lt Leon Lueckenotte located both survivors, but no beacons, radio contact, or other survivors were detected. TSgt Williams and Sgt Stevens were rescued about two hours later by Jolly 32, commanded by Capt Jerry Shipman, using the limited night recovery system (LNRS). Once survivors were on board the HH-53 crew hover-taxied over the crash site looking for other survivors before heading for NKP. Spectre 07, Spectre 12, and Spectre 16 continued to search the area for several hours, but no signs of life were observed. The next day, Royal Laotian Army soldiers were flown to the crash site to search for survivors. The area was teeming with Pathet Lao guerillas, who were advancing on the crash site. After partial human remains were found, a firefight ensued, forcing the withdrawal of the search team. An aerial search continued until noon on 24 December, but no other signs of life were discovered.

> "During a search of the crash site, FAR soldiers noted blood stains, but found no survivors. They recovered a hand from one crew member which was forwarded to USAF

> Hospital Clark AB Philippines for possible identification. On 23 Dec, friendly forces were withdrawn from the area due to security. The search area covered one-mile radius by the ground team and ten miles by air. On 24 Dec, 72 at 1200 hours (local) the formal search and rescue was terminated due to no further contact." AF Form 484, Missing Person Supplementary Report, 2 January, 73 (extract declassified).[65]

The loss of Spectre 17 at 7,500 feet was puzzling. Though a 7,500 ft altitude was slightly riskier than *Thor's* usual 10,000 ft., other pilots had done the same thing in the past to fire the 20mm guns. What wasn't known at the time, the PAVN had positioned a gun on top of a hill to shorten engagement distance and Thor unknowingly flew right over it.

> "The 593rd AAA Regiment headquarters selected a hill a little more than ten kilometers from the Ford 22 choke-point and secretly established a 37mm AAA position on the hill. One evening in December 1972, following the enemy's usual pattern, an AC-130 descended to low altitude to look for targets along Route 22 before it climbed

65 The remains were positively identified via fingerprints as belonging to Capt Joel Birch, Spectre 17 Electronic Warfare Officer.

> back to higher altitude for its flight over the [Ford 22] choke-point. The AC-130 flew straight toward our secret firing position, looking 'as big as a ship' to our gunners." Tien Phong Newspaper, 20 Dec 2012, The Ho Chi Minh Trail's Air Defense Troops and Memories of 1972.

By 23 December, most AC-130 taskings had shifted from *Steel Tiger* VR sectors to the Saigon defense mission, which was one of the highest priorities for US forces. By then, PAVN forces had all but surrounded Saigon, and cease-fire negotiations were ongoing. On Christmas Eve and New Year's Eve, eight of the ten AC-130 gunships launched flew the Saigon Defense mission.

> "The importance of keeping Saigon secure became greater this quarter as cease fire negotiations became more intense. Both Communists and Allies increased their efforts to control as much territory as they could in order to get in as good a bargaining position as possible. The city of Saigon, as the political and psychological center of South Vietnam, became even more important in this territorial-political struggle as the enemy attempted to frighten or discredit the Thieu government." 16 SOS History Records, Oct-Dec, 1972.

Two significant Saigon defense missions were flown on Christmas day. Spectre 18, aircraft 046, commanded by Maj Dougherty, spotted twenty-five vehicles twenty miles northeast of Tay Ninh city, but fire-control issues resulted in only two damaged. *Iron Maiden* (call sign Spectre 09), commanded by Capt McMillen used their 105mm gun to systematically engage armored vehicles poised to attack a friendly position. Eight armored vehicles were destroyed, one was damaged and the attack was repelled.

Heavy use was taking its toll on aircraft and guns. On 26 January, *Jaws of Death* (call sign Spectre 01), commanded by Capt Johnson, located fifteen trucks in *Steel Tiger East*. After working several targets, one of their 20mm guns blew apart while firing. After inspecting the aircraft and finding no damage, the crew continued their mission, ultimately destroying eight trucks and damaging seven. *Pave Spectre* fire control computer failure was also an issue, however, *Iron Maiden* (call sign Spectre 05) flew a successful AC-130E mission on 15 January. This mission, commanded by Capt Ellis, located twenty-one enemy trucks in *Steel Tiger East* ultimately destroying thirteen and damaging two while receiving seventy-nine rounds of 37mm AAA. Two days later, *Bad Company* (call sign Spectre 16), commanded by Capt Lumsden, located two armored personnel carriers, a truck park, and an enemy encampment two thousand meters to the northwest. Teaming with their fighter escorts, they damaged one APC, destroyed eight more, and damaged five trucks.

The Vietnam cease-fire went into effect at midnight Zulu time on 27 January, 1973. From that point on, one AC-130 was placed on alert at Tan Son Nhut Airfield from 1700 to 0600 daily with a new aircraft and crew rotating to and from Ubon every day. The alert aircraft was loaded, fueled and ready to take off at all times while in radio contact with their controlling agency—Blue Chip—for the duration of their alert commitment. This alert posture continued for several weeks.

In Feb, the first CROC configured AC-130-E/H (*Wicked Wanda*) was picked up at Robins AFB, Georgia by Capt Charles Holland and his crew. The aircraft was rigged for pressurization with the crew entrance door installed, gun barrels removed, and closeout panels over the gun ports. Aerospace engineers were convinced the aircraft would not have difficulty climbing to altitude, but actual flight proved otherwise.

> "We found out the computed drag data was quite low for the actual performance of the aircraft. We initially filed for a cruise altitude of 18,000 feet with steep climbs approximately every two hours. We had to level at 16,000 feet since the aircraft had ceased climbing at normal climb power. Our actual flight time from Robins to McClellan AFB was 8.5 hours and we finally made it to 18,000 feet after about six hours of flight time." **CMSgt (Ret.) Gene Eller,** flight engineer.

Wicked Wanda arrived at Ubon on 5 February 1973. Crews welcomed the new gunship with reservations. It was incrementally more capable than the AC-130A but was heavier and less maneuverable. It was designed to be flown "buttoned up" with doors closed. Large Plexiglas blisters were installed in the aft door and at the right scanner station. Many considered it unwise to fly with the doors closed since emergency egress was blocked. Compared to the legacy fleet, the most obvious aspect of the CROC AC-130E/H was the experimental paint scheme. Instead of the usual black fuselage with camouflaged top, *Wicked Wanda* was painted gray with yellow markings. When *Predator* arrived a few days later, it was painted gray with red markings. *Death Angel* sported an experimental almost purple paint scheme crewmen called the Purple Gomer Eater.[66]

During the month of February, mission focus was redirected to TICs in Laos and Cambodia to support Cambodian forces fighting the Khmer Rouge. A hunter-killer team of AC-130s and fighters began to fly missions in Cambodia during daylight hours. After less than a week, daytime results proved mediocre, so the concept was abandoned and AC-130s returned to their nighttime flying role.[67]

On 22 February, the Laotian cease-fire went into effect, but AC-130 missions continued in Cambodia. During this time, the 16 SOS was targeted with a massive propaganda

66 After field trials, a medium gray called "gunship gray" was selected as the standard paint. The gray shaded blended in well with the near-constant light gray overcast that hung over Laos.

67 16 SOS history archives, Jan-Mar, 1973, P 19

campaign by the North Vietnamese, who wanted the US out of Cambodia.[68] Nevertheless, gunship crews continued to fly armed reconnaissance and TIC missions seven days a week in the *Freedom Deal East/West* areas northeast of Phnom Penh. Combined efforts of both AC-130A and AC-130E/H resulted in locating 316 trucks, destroyed 181, and damaged 86. Crews attacked 63 boats, destroyed 45, and damaged 14. They also performed nine TIC missions killing a total of 473 enemy troops, a higher number of enemy KIAs than earlier interdiction campaigns. [69]

> "After the war ended in Laos there were so many aircraft flying missions into Cambodia that the air space became saturated at all altitude levels with the B52's up high the Fast- movers (F-4's, A7, etc.) at mid-height, then Spectre and helicopters at low level that ABCCC had a difficult time coordinating. B-52's were dropping too close to other operations and off target as well. Most aircraft including Spectre started turning nav lights on where we would normally be black. The ground combat controllers of course only wanted the "Big Gun" Spectre." **Sgt Larry Peterson,** Gunner.

---

68 16 SOS History Archives, combat operations, Apr-Jun 1973, p-8, (extract declassified)

69 16 SOS historical archives, Apr-Jun 1973, Mission Recap, 10 May, 1973 (unclassified)

By late April, the dry season ended and rainy weather became a source of frustration when working with troops and convoys.[70] Rules of engagement became more restrictive, and significant delays validating targets resulted in crews waiting for long periods for clearance to fire. On 29 April, Spectre 15 had to wait for one hour and eight minutes for target validation leaving only about ten minutes to fire on targets before bingo fuel. Likewise, on 1 May, Spectre 05 had to wait one hour and twenty minutes for validation. There were also problems with the method of requesting AC-130 support since there was no standard method used by Cambodian ground commanders. This confusion led to miscommunications and enemy force intrusion onto communications frequencies. To counter these problems, a challenge-response checklist was developed to authenticate friendly ground commanders. Because of terrain and foliage, determining BDA was a problem when working TIC missions in Cambodia. Literally firing into the blind, aircrews were unaware of what they had accomplished on the ground. Complicating the problem was the fact that neither the ground commander nor gunship crew were able to see enemy casualties. Furthermore, post-mission daylight reconnaissance photos could not verify damage levels or KIA of enemy personnel in dense vegetation.

BDA reporting was wildly out of proportion in Cambodia since ground commanders estimated the damage level, not actual damage. Perhaps the most accomplished

70 16 SOS History Archives, Apr-Jun 73, page 16 (extract declassified)

inflator/estimator was a ground commander (call sign Hotel Takeo) at the town of Takeo, south of Phnom Penh. On 3 April, Hotel Takeo gave Spectre 08 an estimate of eighty KIA, one machine gun destroyed, one 79mm rocket launcher destroyed and one 57mm cannon destroyed. On 12 April, he estimated Spectre 01 destroyed two 60mm mortars, three machine guns, and killed twenty-five enemy personnel. The next day, he credited Spectre 07 with one 60mm mortar, one 12.7mm machine gun, and seventy KIA. On 14 April, he credited Spectre 06 with fifty-sixty KIA. Then again on 16 April, he credited Spectre 03 with two B-40 rocket launchers, two 30mm mortars, four 79mm grenade launchers, five bunkers damaged, and twenty-thirty KIA.[71] Since BDA was not verifiable, either the town of Takeo was the hottest spot in Cambodia or the ground commander was exaggerating BDA to the point of ridiculousness, but proof of either was never attained.

By late May, AC-130 missions continued, but targets were scarce. The skies over Cambodia became crowded with a high number of strike aircraft. On 1 June, Capt Hampleman's crew (call sign Spectre 06) was flying *Raid-Kills 'Em Dead* when a string of bombs about a mile long exploded about a mile off their right wing. On 24 June, *Iron Maiden* commanded Capt Phillips (call sign Spectre 07), was engaged by three SA-7s while working for Hotel Sac Sam Peony, southwest of Phnom Penh. Defensive maneuvers and deployment of decoy flares proved effective, but one missile

71 16 SOS History Archives, Apr-Jun 73, page 19 (extract declassified)

exploded near the #1 engine requiring it to be shut down. The crew successfully recovered at Ubon.

On 2 June, Spectre 07 discovered forty-nine boats while en route to a TIC. The boats ranged in length from twenty to fifty feet. and were located on the Tonle Sap River. The boats were validated as targets and the crew destroyed eighteen and damaged thirty-one.[72] *Spectre* crews continued to fly missions during July with some level of success, but rainy weather impacted overall mission success. Boats remained primary targets since few land vehicles could be detected operating on the flooded trails.

A significant milestone was achieved on 6 July, when *Jaws of Death* parked after landing, validating one thousand consecutive abort-free AC-130 combat missions. This was a testament to crews and maintainers' outstanding achievement and mission accomplishment. Missions continued unabated through the rainy season. Crews often searched for boat traffic on the Mekong River while en route to their working areas, and boats were plentiful in Cambodia. On 8 August, Spectre 20, commanded by Maj Bill Kehler, searched the Tonle Sap River, then proceeded down the Mekong en route to a TIC mission. Sensor operators located seven barges, each approximately fifty feet long, and requested permission to fire. *Moonbeam* validated the targets, then Spectre 20 went to work, ultimately sinking them all. On August 9, Spectre 01 destroyed a fifty-foot motorized boat and three twenty-foot motorized

72 16 SOS History Archives, Apr-Jun 73, page 15 (extract declassified)

boats. Later on, the same night, Spectre 08 destroyed eight boats. On 12 August, *Grave Digger* (call sign Spectre 20), commanded by Maj Gates, destroyed five boats on their way to a TIC mission and three more on their way back. Finally, Spectre 06, commanded by Col Harry W. Williamson, landed aircraft 69-6574 at 0805 local time on 15 August. This was the last AC-130 combat mission flown during the Vietnam War.[73] Coinciding with the last combat mission flown was a record of 1,327 consecutive missions without an abort.

The crew of Spectre 06 after completing the last AC-130 combat mission of the war. (USAF Photo)

73 The crew of Spectre 06 included; Col Harry Williamson, Majs Paul Stenback, David Davis, and Peter Traversa, Capts Robert Conners, Raymond Honacker, and Cecil Weaver Jr., 1Lt's Bari Matherne, Harry Hudson, and Gene Sawyer, 2nd Lt Robert Allmon, SMSgts Stanley Noret, and Frederick Brenner, TSgts Roderick Gibbs Jr., and Robert Potts, SSgt William Krallis, Sgts Michael Tesch, and Lawrence Ericson.

Once combat operations halted, crewmen were routinely assigned "busy work" by the 8th TFW including base maintenance details while Air Staff decided the fate of the AC-130. Local missions were flown solely to maintain basic proficiency and efforts focused on support of Search And Rescue (SAR) missions. It didn't take long for Air Staff to start paring down officer positions. The USAF had a shortage of navigators and since there were five navigators (NAV, FCO, EWO, IR, and LLLTV) on each AC-130 crew, the 16 SOS was targeted for easy pickings. Discussions began to convert two navigator sensor operator positions to enlisted positions. The shooting war was clearly over for the 16 SOS, but the 7th Air Force kept the squadron in place at Ubon as part of the overall SAR effort.

On 27 November 1973, Spectre 65 was involved in a real-world SAR effort that played out the exact scenario they had trained for. While flying a routine training mission, Capt Tom Hebblewhite's crew was diverted to a crash site of an A-7 south of Ubon. While there, the crew assumed the role of on-scene commander working with rescue aircraft and ultimately escorted the rescue helicopter back to Ubon[74].

74 16 SOS Historical Archives, Oct-Dec, 1973, P 7 (extract declassified)

MISSION DEBRIEF: 1968–1973

The 7th Air Force claims to have destroyed more than 13,000 trucks while PAVN records indicate 11,347 were destroyed. Of these, the vast majority were destroyed by *Spectre* crews. Though the actual number of trucks or tons of supplies destroyed may never be known, there is no doubt *Group 559* regarded the AC-130 as their archenemy and took exceptional measures to counter *Spectre*. The fact that a converted cargo aircraft was the most effective truck killer during every phase of *Commando Hunt* surprised PACAF leadership, tacticians, and strategists. –The AC-130 bridged the TIC capability of the AC-47 and AC-119 and was used with great effect during many TIC scenarios, especially during the Battle of An Loc.

The average number of combat missions flown per air-crew member during a one-year tour ranged from eighty to one hundred, but those who served for multiple tours may have flown three hundred or more combat missions. From 1968 through 1973, gunship personnel were awarded numerous Silver Star Medals, Distinguished Flying Crosses, and Air Medals, for valor and meritorious achievement and those wounded or killed in action were awarded the Purple

Heart medal. Regardless of threat level, *Spectre's* commitment to interdict the PAVN logistic system never wavered.

> "Looking back, seems to me we were naive about threats we encountered on the trail and analyzing what we did know about NVA AAA and SAM experiences ... Intelligence community offered very little help. In hindsight, I don't know what we could have done to prevent Gunship losses we incurred, but think we were very lucky not to have lost more aircraft now knowing what massively heavy AAA defenses we were up against." **Col (Ret.) Jim Kyle,** Spectre pilot and operations officer.

> "I feel as squadron EWOs (BC) we as a group did not do enough to ensure crews were knowledgeable of the threat, threat electronics, our threat countermeasures capability, and tactics. Knowledge and actions were minimal and haphazard at best. There were a lot of misconceptions about our ECM equipment and expendables (chaff and flares.) Hindsight is a great eye-opener. "**Maj (Ret.) Henry "Hank" Rother,** Spectre EWO

Thousands of missions were flown in support of US and allied ground forces and against *Group 559*. *People's Army of Vietnam* records indicate ten thousand North Vietnamese soldiers and workers were either killed or died of accidents or disease while building, operating, or maintaining the Ho Chi Minh Trail. The sheer number of trucks operating on the Ho Chi Minh Trail every night surpassed the ability of AC-130 and AC-119K crews to locate, target, and destroy them and their cargo. Regardless of the high level of effort by *Spectre* crews, the Ho Chi Minh Trail remained open throughout the war. When the *K-Trail* opened in 1972, large convoys were able to transit nearly the entire length of the trail undetected, even by AC-130 sensors.

Some of the most telling aspects of the effectiveness of the AC-130 do not come from gunship personnel or those they supported. Instead, it comes from members of the North Vietnamese Army who endured nearly six years of constant attacks by *Spectre* crews.

> "Over the Ho Chi Minh Trail, AC-130E aircraft equipped with sensors to locate targets and 40mm rockets (sic) to destroy these targets, flew constantly throughout the night.[75] They destroyed many of our trucks. Our combat fighters who drove the trucks courageously, resolutely, and cleverly overcame horrendous obstacles to bring supplies to the front lines, and some comrades sacrificed their lives in the cabs of their trucks. At the same time, faced with the blood and fire of

75 The AC-130E was not employed until the 1971-1972 dry season. It appears most PAVN records indicate AC-130E as a general reference to all AC-130 models.

> combat on the supply route to the South, some drivers wavered, and there were even some who abandoned their vehicles even when they heard just the sound of the enemy's AC-130 gunships flying overhead..."
> Victory in Vietnam_(PAVN History Archives)
> Mr. Merle L. Pribbenow, University Press, Kansas 2002.

From 1968-1973, *Spectre* combat operations in Vietnam, Laos, and Cambodia resulted in countless lives saved during close air support missions and thousands of tons of PAVN supplies destroyed. These achievements came with a heavy price with the loss of six AC-130 gunships and fifty-two crewmen.

Aircraft and personnel losses during Southeast Asia combat operations	
Carter, 54-1629 The Arbitrator (5/24/1969)	Spectre 13, 55-0044 Prometheus (3/29/1972)
SSgt Cecil Taylor	Maj Irving Ramsower
SSgt Jack Troglen	Maj Harry Brauner
Adlib 11, 54-1625 The Warlord (4/21/1970)	Maj Howard Stephenson
Lt Col Charles Rowley	Capt Barclay Young
Li Col Charles Davis	Capt Richard Halpin
Maj William Brooks	Capt Richard Castillo
Maj Don Fisher	Capt Curtis Miller
1stLt John Towle	Capt Charles Wanzell III

MSgt Robert Ireland	MSgt Merlyn Paulson
Sgt Ronnie Hensley	SSgt James Caniford
Sgt Thomas Adachi	SSgt Edwin "Jack" Pearce
Sgt Stephen Harris	SSgt Edward Smith Jr.
AIC Donald (Mike) Lint	Sgt William Todd
	AIC Robert Simmons
Spectre 11, 55-0043 Odin (6/18/1972)	Spectre 17 Aircraft 56-0490 Thor (12/21/1972)
Maj Robert Harrison	Maj Paul Meder[vi]
Maj Gerald Ayers	Maj Francis Walsh Jr[vii]
Capt Mark Danielson	Capt Harry Lagerwall
Capt Paul Gilbert	Capt Stanley Kroboth
Capt Robert Wilson	Capt Thomas Hart III
MSgt Jacob Mercer	Capt Robert Liles Jr.
TSgt Richard Cole	Capt Delma Dickens
TSgt Donald Klinke	Capt Joel Birch
SSgt Larry Newman	2nd Lt George McDonald
SSgt Richard Nyhof	TSgt James Fuller
Sgt Stanley Lehrke	TSgt John Winningham[viii]
Sgt Leon Hunt	Sgt Robert Elliot
	AIC Charles Fenter
	AIC Rollie Reaid
Aircraft 69-6571 The Destroyer (3/30/1972)	
Controlled bailout, all were rescued	

vi. Was student IR Operator onboard for an orientation flight
vii. Was an F-4 pilot onboard for an orientation flight
viii. Was a Standardization/Evaluation Illuminator Operator administering a check ride to Sgt Carl (Ned) Stevens, who survived bailout.

SITUATION REPORT: 1974–1975

When combat operations ended, Col Hopingardner, 16 SOS commander, set out to make the squadrons' unique capabilities known throughout PACAF. This traveling roadshow centered on the AC-130's ability to perform search and rescue, command and control, and fire support. Two gunships were sent TDY to Osan, Korea to put on a firepower demonstration for high-ranking South Korean officers.

The overall situation at Ubon was quiet while crews and maintainers were kept on alert status, but rarely launched. In the summer of 1974 Ubon was scheduled to close and the 16 SOS was reassigned to Korat Royal Thai Air Base. By the fall of 1974, seventeen aircraft and 435 people moved to Korat, assigned to the 388th Tactical Fighter Wing.

As the SEA drawdown continued, pilferage from local criminals increased, especially at night. Thieves were stealing everything they could get their hands on, even the barbed wire protecting the perimeter of bases. Base Exchange trucks moving merchandise from the port on the Gulf of Siam to the bases at Korat and Udorn were hijacked. Soon after the problem was realized, *Spectre* was tasked to provide nighttime security for bases at Udorn, Nakhon Phanom, and Korat. When a crew detected suspicious activity, they radioed Security Police and gave them vectors to the site. Along the way, someone discovered putting reflective tape (later called GLINT tape) on the top of a Base Exchange truck's cab enabled the LLLTV operator to used the GLINT to find stolen trucks, even when in heavy traffic. Glint tape had become the new standard for marking friendly vehicles.

In March 1975, the Thai Parliament elected Prime Minister Khurkrit Parmot who demanded all US forces leave Thailand by the spring of 1976. There was initial confusion and debate as to where the gunships would go. There was talk of sending all AC-130A's directly to the aircraft boneyard at Davis Monthan AFB, AZ while assigning all AC-130H's to the Air Force Reserves.[76] The final plan transferred all H-models to Tactical Air Command at Hurlburt Field, FL, and all A-models to the Air Force Reserve at Duke Field FL.

In early April 1975, seven AC-130 crews were put on alert when the US-backed Khmer Republic was overthrown by the communist Khmer Rouge. Beginning on 12 April, AC-130s began flying airborne alert missions for *Operation Eagle Pull*, the evacuation of Phnom Penh, Cambodia. When the evacuation was complete, not a single round was fired and all mission objectives were met.[77]

76 The "Boneyard" is jargon for the dry aircraft storage facility near Davis-Monthan AFB, Arizona.

77 U.S. Air Force, History of the 388th Tactical Fighter Wing, 1 April 1975-30 June 1975, Vol II,(S), (extract declassified)

On 29 April, *Operation Frequent Wind*--the evacuation of Saigon, Vietnam—began and AC-130 gunships were tasked to provide fire support on request. Because it was not possible to land in Vietnam for fuel, gunships were equipped with external fuel tanks to extend time over target. During these missions, crews orbited downtown Saigon at 14,000-14,500 ft. to avoid possible anti-aircraft fire and SA-7 missiles. During one mission, a hunter-killer flight of AC-130s and F-4s encountered a 57mm AAA site delivering accurate fire, so the site was first marked by the gunship crew then destroyed by F-4D Phantoms.[78] VNAF AC-119s provided fire support and one of them was shot down.

Thousands of people were evacuated by fixed-wing aircraft, but the last phases were done by helicopters moving refugees to ships offshore. Dozens of helicopters were flying over the city, many of them without lights because of the infrared (IR) missile threat. Spectre 71 remained orbiting on station during the morning until the last helicopter left. Overall, gunship crews flew eight sorties for a total of 46.9 flight hours during the evacuation operation. By the end of April 1975, communist forces had a firm grasp on Laos, Vietnam, and Cambodia. These governments capitalized on the US withdrawal from the region, going to great lengths to harass personnel and stymie US interests in the region. The recently established Cambodian Khmer Rouge government was exuberant with their newly found power and may have felt invincible, at least until 13 May 1975.

78 Ibid

By early summer of 1975, the AC-130A fleet began to depart Korat for Duke Field, Florida. After reaching the United States, they reconstituted as the 711 SOS reporting to the 919 Special Operations Group, Air Force Reserve. These gunships arrived in Florida in pretty weary shape. Two were deemed economically irreparable for corrosion and retired to Davis Monthan, AZ leaving ten AC-130A in service.[79] On 30 June 1975, Hurlburt's 415 SOTS was deactivated and its two AC-130H gunships (69-6567 and 69-6568) were reassigned to the 8th SOS, Hurlburt Field, Florida.

79 History, 919 Special Operations Group, 1 Jan 1975-30 June 1976, (U) 12-16.

CHAPTER THREE

RECOVERY OF THE SS MAYAGUEZ

SS Mayaguez, National Museum of the US Air Force Photo

On the afternoon of 12 May 1975, Khmer Rouge gunboats intercepted the US merchant ship *SS Mayaguez* and its crew near the Cambodian island of Paulo Wai. Though the ship was transporting consumer goods for the PX, a Khmer Rouge officer believed the ship was transporting military weapons. The ship's captain, Charles Miller, was directed to set a course for the Cambodian port of Kompong Som, some ninety miles away. The skipper realized his ship and commercial cargo would be gone forever if it reached the mainland, so he convinced his captors the ship's radar was defective and he could not safely bring the ship to port in the darkness. Accepting the captain's ruse, Khmer Rouge officer directed the *Mayaguez* to drop anchor for the night. The next morning, Captain Miller was ordered to proceed, but he stalled as long as he could with the hope that a rescue force was on the way. Unsure of the ship's status, Khmer Rouge officer directed the *Mayaguez* to be sailed within a mile and a half of the shoreline and drop anchor near the Cambodian island of Koh Tang.

The first U.S. military action was aerial and naval reconnaissance intended to locate the *Mayaguez* and its crew. At 1600 on 13 May, a Khmer Rouge official ordered the *Mayaguez* crew to board two fishing boats tied up alongside the *Mayaguez*. A US Navy P-3 *Orion* maritime patrol aircraft reported more than thirty people boarding the fishing boats. The boats were then escorted by several Khmer Rouge gunboats to a small cove near Koh Tang. At about the same time, President Ford, through the Commander-in-Chief Pacific (CINCPAC), code name Blue Chip, alerted the 16 SOS to provide fire support for a yet to be defined rescue and recovery operation.

> "Our job was to be ready to maintain surveillance and halt waterborne traffic through the night, with no idea how many flights would

> be involved. It took over an hour to fly down to Koh Tang and each aircraft could remain on scene for about four hours before coming back. Neither the AC-130A nor AC-130H were capable of aerial refueling. The gunship was, therefore, a limited resource and would have to be conserved if it was to be the sole source of surveillance and firepower in the hours of darkness." **Lt Col David Mets,** 16th SOS Operations Officer.

Fighters and reconnaissance aircraft overflew the *Mayaguez* and searched for signs of the crew during daylight while AC-130 gunships were tasked with night missions to take advantage of the unique capabilities of our infrared (IR) and low light level television (LLLTV) systems. The gunships also carried LUU-2 flares to assist fighters and forward air controllers (FACs) when needed.

The first AC-130, Spectre 11, took off from Korat Royal Thai Air Force Base, at 1710 hours, commanded by Capt Edward Burns III. Upon arrival at the working area, Spectre 11 contacted the EC-130 Airborne Battlefield Command, Control, and Communications (ABCCC) aircraft, call sign Cricket, for instructions. Cricket directed them to search for boat traffic with their IR and LLLTV sensors. The *Mayaguez,* code name Mother Ship, appeared dead in the water with no activity onboard. Spectre 11 detected the cluster of Khmer Rouge patrol boats surrounding the two fishing boats in the Koh Tang cove and patrol boats shuttling supplies from the cove to an encampment near the center of the eastern beach.

After about thirty minutes, Cricket directed Spectre 11 to two islands about fifteen miles south of Koh Tang. Activity on both islands appeared to be quiet, so they returned to Koh Tang. After rolling into orbit at 6500 feet, Spectre 11 began to receive .50 caliber fire that fell short of their altitude. Though the gunship had the gun in their sights, they were not authorized to return fire since it was unknown where the *Mayaguez* crew was being held. Cricket redirected the gunship back to the *Mayaguez* where it orbited for about forty-five minutes, monitoring boat activity in the cove.

When Spectre 11 ran low on fuel, Spectre 21, commanded by Capt Michael Mueller, arrived to continue surveillance operations. The crew remained overhead for three hours and twenty minutes, locating and reporting multiple gun sites in the area surrounding the Khmer Rouge encampment. Gunship crews noted increased boat traffic as the night wore on. When Spectre 21 ran low on fuel, Spectre 31 replaced them.

Gunship crews flew blacked out to limit acquisition by AAA gunners. When an RF-4 Phantom II tactical reconnaissance aircraft unexpectedly flew over the island dispensing photoflash cartridges, it was more than a little surprising.

> "I was hanging off the ramp when out of nowhere an RF-4 flew from 7 to 4 o'clock shooting out a line of bright flares. It being

> pitch dark and not knowing they were coming scared the crap out of me. He was close." **SSgt Joe Percivalle,** Spectre 31 Illuminator Operator.

At about 0400 on 14 May, Spectre 41 detected a patrol boat leaving the cove, heading towards the mainland. Cricket cleared the gunship to fire warning shots in front of the boat to stop it or cause it to turn around. The gunship fired 40mm rounds in front of the boat, every time it headed for the mainland. Ultimately the patrol boat appeared to run aground.

Just before sunrise, CINCPAC authorized Spectre 51, commanded by Capt Gregory Kirkland, to fire warning shots near three patrol boats attempting to leave the area. Fires from the gunship and F-111s turned two boats around, but the third continued towards that Cambodian mainland. A-7 Corsair IIs dropped tear gas in front of the fleeing boat, but it continued towards the mainland. Eventually, it was strafed and sunk by an A-7 while the other two secured alongside the *Mayaguez*.

Later that morning, a loaded fishing boat left the cove heading towards the Cambodian mainland. Aircrews reported a number of people huddled into the bow of the vessel and there were concerns the people could be the *Mayaguez* crew. Fighters strafed, bombed, and dropped CS gas in front of the boat in an attempt to stop it, but the boat evaded fire and continued on course. By 1000 hours, the boat successfully made it to Kampong Som. It was unclear if the fishing boat transported the *Mayaguez* crew. Without confirmation of crew whereabouts, the US assumed there was still a strong possibility some, if not all, were held captive on Koh Tang.

By mid-day on the 14th, the operation entered a new phase that allowed direct strikes on gunboats, or any boat that did not appear to be capable of holding passengers. Fighters bombed and strafed boats, sinking or damaging several. Near midnight, Spectre 41, commanded by Capt Henry Markulis, observed a boat heading towards the *Mayaguez* and was cleared to fire on it. Spectre 41 fired 192 rounds of 40mm, stopping the gunboat only 200 yards from the *Mayaguez*. A few hours later, Spectre 61, commanded by Major J. Cobble, replaced Spectre 41, performed overhead surveillance until low on fuel, then diverted to U-Tapao to refuel before returning to Koh Tang.

As the situation evolved, the President made a decision to mount a full-scale military operation to rescue the *Mayaguez* crew by force. Fire support for the operation would be provided by USAF fighters: F-4s, A-7s, and F-111s, and by AC-130 gunships. Forward air control was to be provided by the A-7 flight leads. The guided-missile destroyer, *USS Henry B. Wilson,* call sign Black Velvet, would also provide fire support once it arrived in the battle area.

The rapid pace of events, unreliable intelligence, and overlapping chains of command extending to the White House caused more than a few problems. Working with extremely limited tactical information and no exact confirmation of the *Mayaguez* crew's location, mission planners assumed Khmer Rouge guards would put up only token resistance before releasing the crew. Also unknown to US mission planners and undetectable by aerial reconnaissance methods, was that the landing zones (LZs) on East and West Beaches bracketed a heavily fortified encampment of well-trained and dug-in Khmer Rouge soldiers.

By the third day, Khmer Rouge commanders at Phnom Penh realized the situation had rapidly grown out of hand. They wanted to end it, but since there were no diplomatic relations established with the US, they could not communicate their intentions. At the same time, Captain Miller had established a dialogue through an English-speaking interpreter with the commander at Kampong Som. The commander ordered the release of Capt Miller, his chief engineer, and seven other crew members. These individuals were to return to the Mayaguez to light the boilers and radio Bangkok to tell the US to cease military operations. Unfortunately, the decision was reached after dark, so they decided to wait until daylight to sail back to the *Mayaguez*.

During the early morning hours of the 15th, the National Security Council met to discuss options. Since there was no diplomatic response from the Khmer Rouge government, President Ford ordered military action.

At first light on 15 May, Capt Miller and his partial crew boarded a fishing boat to return to the *Mayaguez*. About the same time, 250 Marines boarded 8 CH-53 and HH-53 helicopters to assault Koh Tang. Three more helicopters loaded a boarding party to be delivered to the *USS Holt* to recover the *Mayaguez*. Spectre crews searched both beach LZs and observed the gun sites that had been identified on 13 and 14 May, but the situation appeared quiet. Meanwhile, the first four helicopters approached the island to begin simultaneous landings on the East and West beaches. Knife 21 and Knife 22 approached the western beach without opposition but as Knife 21 touched down, heavy fire erupted from entrenched positions. All twenty Marines on the helicopter exited while under fire.

Landing attempts on the eastern beach were no less dangerous. As Knife 23 and Knife 31 approached the LZ, Khmer Rouge guns remained silent until both helicopters entered a hover, then opened fire. Knife 23 was just a few feet above the HLZ when it was hit, losing its tail rotor. The aircraft spun out of control and crashed onto the beach. Remarkably, twenty Marines led by 2nd Lt John Cicere, a USAF combat photographer, and the entire crew survived the crash. Within seconds of arriving over the LZ, Knife 31's left external fuel tank caught fire, followed almost immediately by a rocket-propelled grenade hit to the cockpit. The helicopter crashed into the surf less than one hundred feet from the shoreline, directly in front of a Khmer Rouge gun position. Remarkably, eighteen of the twenty-six onboard survived the crash but were fired upon as they swam out to sea. Among the survivors was the battalion FAC, 1st Lt Terry Tonkin, who used a survival radio to call for fire support as he swam out to sea. He was eventually rescued by the *USS Wilson* along with twelve other survivors.

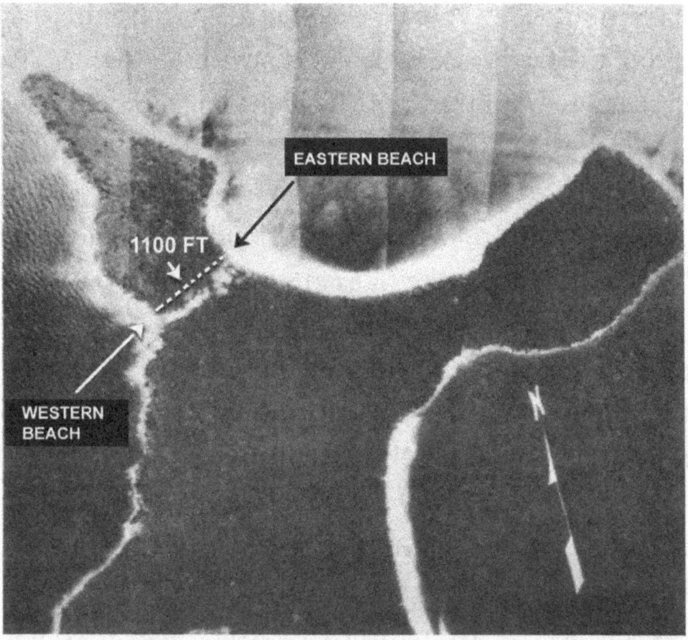

West beach (lower left) and East beach (above, center) landing zones on Koh Tang Island were separated by a short strip of land in which the majority of Khmer Rouge soldiers fought from dug-in fighting positions. The remainder of the island was covered with heavy jungle growth. (DoD Photo)

Mission commanders were surprised by the level of resistance. Since radios had been lost or destroyed during the initial insertions, communications between Marines on the island and the fast FACs were largely ineffective. The A-7s could not accurately pinpoint camouflaged Khmer Rouge positions in the dense jungle. Gunship crews were orbiting nearby in frustration since they had reported the gun sites' locations to intelligence two days earlier. For unknown reasons, the information was not passed to the FACs. Coincidentally, gunship crews had the Khmer Rouge positions in their sights that morning but neither Blue Chip,

Cricket, nor the FAC would clear the AC-130s to fire, instead favoring the fast jets. Tragically, many of the fighter strikes were complete failures with a high percentage of bombs missing the island.

By 1000 on the 14th, the *USS Holt* was towing the *Mayaguez* away from Koh Tang Island. Just after noon, a fishing boat flying white flags approached the *USS Wilson*. The skipper of the *Mayaguez* and his engineers were released and the fishing boat turned around to retrieve the rest of the *Mayaguez* crew. Coincidentally, the captured Thai fishing boat was the same vessel that escaped the gauntlet of fire as it headed for the mainland two days earlier. Although the *Mayaguez* crew had been released, the operation was far from being over.

When Spectre 61 returned to Koh Tang, it contacted Cricket for tasking and was handed over to the Search and Rescue (SAR) coordinator, call sign Rotor, for tasking. Spectre 61 located the enemy that were firing on the survivors and passed the coordinates to Rotor. Even though Spectre 61 was ready to fire, they did not get clearance to fire because there were fighters on the scene. After the fighters had expended their ordnance, Rotor cleared Spectre 61 to fire. After shooting eight to ten rounds of 40mm, the crew was requested to break off because there were more fast movers to bring in.

A short time later, while fighters attempted to identify and hit targets with little success, Spectre 61 was directed to return to base. Five minutes later Spectre 61 was recalled to support two Marine elements which were trying to link up. The two commanders on Koh Tang were separated and did not know each other's location. Plus, they needed the gunship to suppress the enemy ground fire. Cricket cleared Spectre 61 to fire on any targets requested by the Marine commanders.

Spectre 61 put down suppressive fire on enemy positions while Jolly 41 made a low pass over the western LZ. Jolly 41 identified Khmer Rouge positions and passed the locations to the gunship crew. Then, Spectre 61 fired 105mm directly on the heavy machine gun positions. Jolly 41 landed while Spectre 61 continued to lay down suppressive fire.

At 1430, the helicopters attempted to extract personnel stranded on the East beach. Before the helicopters began their approach, an A-7 dropped riot control agent near the beach but a strong wind blew the tear gas out to sea.

With time running out for the remaining Marines on Koh Tang and the *USS Holt* nearby, 1st Lt Robert Blough, aircraft commander of Jolly 44, flew directly to Koh Tang, picked up a load of Marines, and unloaded them on the nearby *USS Holt*. Since the HH-53 was too heavy to land on the destroyer, Blough hovered just above the ship, touching his nose landing gear on deck as the Marines boarded the ship. Jolly 44 then returned to Koh Tang, made several unsuccessful approaches in the smoky darkness, but finally landed while under fire to extract another load of Marines. A short time later, Knife 51 circled the LZ in complete darkness while Capt Seth Wilson, OV-10 FAC, (call sign Nail 68), cycled his aircraft landing lights on and off to guide Knife 51 into the LZ. Meanwhile, Nail 68 kept Spectre 11 in a holding orbit southeast of the island while fighters dropped their ordnance with mixed results. Finally, unsatisfied with fighter strikes, Nail 68 directed Spectre 11 into position over Koh Tang. The gunship crew began firing as directed, initially shooting 105mm high explosive rounds, but the heavy jungle canopy proved to be a challenge. Spectre 11's LLLTV sensor operator, 1st Lt Don Raatz, was unable to clearly see the rounds' impacts because of the dense jungle canopy. After switching to 40mm Misch Metal, they could see the impacts and accurately adjust fire.

Spectre 11 fired its 40mm gun almost continuously after that, with only two short breaks to drop flares. When the storage barrel for expended 40mm brass was full, the crew fired two long bursts of 20mm high explosive incendiary (HEI) rounds while the storage barrel was being emptied. All the while, after three aborted landing attempts, Capt Richard Brims landed Knife 51 on the West Beach while Spectre 11 laid down constant 20mm and 40mm suppressive fire on nearby Khmer Rouge positions.

The intensity of the firefight was such that the Marines, their attention directed outwards at the hostile fire positions barely noticed the newly arrived helicopter. When several seconds passed, and none of the Marines made an effort to disengage and withdraw into the helicopter, TSgt Wayne Fisk, a pararescueman on Knife 51 disembarked and fired several rifle rounds into the air to get their attention. An orderly extraction began until 29 Marines were aboard Knife 21. Unsure if all Marines were accounted for, TSgt Fisk signaled the flight mechanic, SSgt Marion Riley, that he would perform a final sweep of the beachhead. Maneuvering down the beach in the dark, illuminated only by overhead flares and the muzzle flashes, TSgt Fisk was nearly bracketed by advancing Khmer Rouge soldiers. Meanwhile, Spectre 11 unleashed a long burst of 20mm HEI fire.

> "...a magnificently violent, shimmering curtain of red, orange, and yellow death. It began as a single red line descending from my right and grew, like the Aurora Borealis, across the front of me to my left, then it went back and forth. It was like a huge protective curtain being drawn in front of me, and I knew I was safe, but everything in it was dying. I felt omnipotent." **CMSgt (Ret.) Wayne Fisk.**

After Spectre 11's strike, Fisk searched the beachhead for remaining Marines and the body of LCpl Ashton Loney, which he had seen earlier that afternoon. Unable to locate any Marines or Loney, he reboarded Knife 51 and they flew to the *USS Coral Sea* while all AC-130s returned to Korat.

AC-130 mission summary for 13–15 May 1975			
Date	TO time/Duration	Call Sign	Aircraft Commander
13 May	1710/5.8	Spectre 11	Capt Edward Burns III
13 May	2005/5.6	Spectre 21	Capt Robert Hilb

13 May	2158/5.3	Spectre 31	Capt Peter King
13 May	2328/7.2	Spectre 41	UNK
14 May	0137/7.5	Spectre 51	Capt Gregory Kirkland
14 May	1600/5.6	Spectre 11	Lt Col David Mets
14 May	1805/6.5	Spectre 21	Capt Michael Mueller
14 May	2030/5.4	Spectre 31	Capt Robert Hilb
14 May	2150/5.8	Spectre 41	Capt H. Markulis
14 May	2325/6.6	Spectre 51	Capt Edward Perry
15 May	0117/10.3	Spectre 61	Maj John Coble
15 May	0117/5.9	Spectre 11	Lt Col J Daughtry
15 May	1607/5.9	Spectre 21	Capt Terry Talbot
15 May	1905/5.3	Spectre 31	Capt Warren Knouff (CP)
15 May	2025/0.7	Spectre 41	UNK

SS MAYAGUEZ DEBRIEF, 1975

The 16th SOS Commander's After Action Report (AAR) detailed positive and negative aspects of AC-130 involvement in the Mayaguez operation. Gunship crews reported difficulties working with command and control and the FACs in particular. The AAR surmised leadership and mission planners did not know how to effectively employ the gunships.

The recovery of the *Mayaguez* was an operation ideally suited to the special capabilities of the AC-130 gunship. While the gunships did make significant contributions to the mission, it is also true that they could have done a good deal more to ensure success. The 16 SOS was extensively trained for troops in contact situations and Koh Tang Island was almost a classic example of such a situation. While the AC-130s did provide close air support for the assault force and suppressive fire for the helicopters, they also could have helped locate survivors and guide the helicopters to the exact location of the ground forces being extracted.

> "Recommend that a briefing tour be sent to inform the appropriate command and con-

> trol agencies, and especially ABCCC units, on the characteristics of the AC-130 gunship. Recommend that an officer with gunship experience be assigned to battle staffs at all levels. One example where the desired awareness may have helped occurred when three helicopters were lost to AAA fire-and others were damaged. Many AC-130 crews had previously sighted these AAA sites and had been fired upon by them. None of the AC-130s received any battle damage and were invulnerable to that caliber threat when flying in their normal firing orbits. For two nights, before the assault, the crews reportedly had the AAA in their sights, but were denied permission to fire upon them."
> **Col James Wyatt,** Commander, 16 SOS.

The *Mayaguez* operation was considered the last action of the Vietnam War, thus ending the AC-130's combat role in Southeast Asia.

GHOSTRIDERS *MORS DE CAELIS* EPILOGUE, 1968–1975

The AC-130 program was not devised, controlled, or marketed by the commercial aviation industry. Instead, Air Force Systems Command took matters into their own hands to enable rapid development and fielding of a needed combat capability designed for a specific mission. First devised as a temporary weapon system to fill urgent needs, crews and maintainers virtually reinvented the wheel on a regular basis to keep those old aircraft in working order. Without a doubt, the AC-130 program would have failed if not for the extraordinary heroic efforts of those who flew in dangerous air over Laos, Cambodia, and Vietnam. The accomplishments of cadre maintainers and aircrews were especially commendable given the fact they inherited some of the oldest, broken-down C-130As in the inventory. Nevertheless, *Spectre* crews developed tactics, techniques, and procedures for close air support, interdiction, command and control, and search and rescue.

For nearly six years, *Spectre* crewmen braved intense AAA fire, radar SAMs, and SA-7s with little training, guidance, or recognition of actual enemy threat weapon capabilities. Tactics and procedures were developed on the fly as a result of observation of enemy activity, battle damage, and loss of aircraft and fellow crewmen. Maintainers worked long hours to keep aircraft in operational condition, many times resorting to nonstandard and resourceful appropriations.

On 9 December 1975, all AC-130H gunships based at Korat had their guns removed for their long trip to Hurlburt Field, FL. The 388 Tactical Fighter Wing Commander presided over a sendoff ceremony for the last gunship in Southeast Asia. It was the eve of the American bicentennial celebration. Thus, the Wing had deliberately held back AC-130H 69-6576, painted the numbers 576 over the radome in red, white, and blue, and painted the phrase The Spirit of 76 underneath the cockpit windows. Col Neil Eddins handed the squadron guidon to the aircraft commander, and the crew scrambled aboard for the last flight from Korat. Passing over the field after takeoff, the crew waved its wings in farewell.

The AC-130 gunship legacy continues today with new aircraft and aircrews who consistently put the right effect, in the right place, at the right time. Either by design or by chance, the AC-130 gunship certainly has earned a unique position in military history.

EVENT CHRONOLOGY

1968

Feb, Detachment 2, 14th Air Commando Wing, deploys to Ubon, Thailand with *Gunship II.*

14 Jun: *Gunship II* deployed to Tan Son Nhut, Vietnam to counter rocket threats to camps and airfields.

28 July: *Gunship II* destroyed at least five trucks while searching Route 246 in Tay Ninh Provence.

18 Aug:*Gunship II* provided TIC support at Katum Special Forces camp.

25 Sep: *Gunship II* defended Thien Ngon Special Forces Camp, preventing overrun.

3–8 Oct: *Gunship II* completed four TIC missions at Thuong Duc, Vietnam, destroying six buildings.

31 Oct: The 16th Special Operations Squadron *"Spectre"* was activated under the command of Lt Col G. S. Fink.

12 Nov: *Gunship II* flew its last combat mission before reassignment to CONUS.

Nov: *Mors De Caelis, Gomer Grinder* and *The Arbitrator* arrived on station at Ubon.

24 Nov: Operation *Commando Hunt* began.

14 Dec: *The First Lady* arrived at Ubon.

Dec: Spectre flew its first mission with F-4 escorts from the 497th Tactical Fighter Squadron "Nite Owls."

1969

17 Jan: ASD Maintenance Assistance Team arrived at Ubon to initiate modification program.

4 Feb: *The First Lady* destroyed 10 boats, 4 trucks.

6 Feb: *Mors De Caelis* defended Laotian outpost at Ben Thateng against NVA troops.

22 Feb: *Mors De Caelis* destroyed 25 trucks on Laotian Routes 12, 128, 1281.

23 Feb: *Mors De Caelis* destroyed 23 trucks along Laotian Routes 12, 128, 1281.

3 Mar: *Mors De Caelis* received battle damage from 37mm AAA.

5 Mar: *The Arbitrator* received battle damage from 37mm AAA.

9 Mar: *The First Lady* destroyed 21 of 34 trucks sighted along Laotian Route 12c.

23 Mar: Lt Col George Liotis assumed command of the 16 SOS from Lt Col Garvie Fink.

11 Mar: *Gomer Grinder* destroyed 11 of 13 trucks sighted along Laotian Routes 92c and 92e.

16 Mar: *The Arbitrator* destroyed 24 of 40 trucks sighted on Laotian Route 8 near Delta 75.

18 Mar: *The Arbitrator* diverted to a troops in contact mission in Laos.

21 Mar: *The First Lady* destroyed 15 of 23 trucks seen along Laotian Routes 914, 234.

27 Mar: *The First Lady* struck by 23mm AAA while working along Laotian route 92d.

1 Apr: Squadron tactical call sign "Spectre" discontinued by order of BGen Bevan.

3 Apr: *Mors De Caelis* received battle damage from 37mm AAA.

7 Apr: *Gomer Grinder* kills all 25 trucks found; the first 100% mission kill rate.

8 May: *The Arbitrator* engaged suspected enemy helicopter.

24 May: *The Arbitrator* hit by 37mm AAA resulting in a crash landing.

29 May: *Gomer Grinder* struck by enemy AAA fire and landed at Nakhon Phanom.

25–28 Jun: Four missions flown in support of Lima Site 108 located at Muong Soui in Northern Laos.

30 Jun: *Mors De Caelis* located and destroyed enemy storage area in southern Laos.

8 Jul: *Ghost Rider* experiences "short round" incident while supporting TIC mission in Southern Laos.

23 Jul: *Ghost Rider* severely damaged by fire in the left wing.

28 Jul: Ubon RTAB attacked. AC-130 launched to provide flare support.

29 Jul: Lt Col Simone replaced Lt Col Liotis as 16 SOS Commander.

31 Jul: *Mors De Caelis* has "*Black Crow*" system installed for test & evaluation.

17 Aug: *Mors De Caelis* kills five trucks using IR tracking.

18 Aug: *Mors De Caelis* kills nine vehicles.

1 Oct: Significant TIC mission flown by *Mors De Caelis* resulting in 100 KBA.

10 Oct: *Ghost Rider* completes first significant dry season mission in *Steel Tiger* East.

29 Oct: Two additional AC-130A gunships arrive at Ubon (six on station 623, 628, 129, 630, 627, 625).

6 Nov: *Gomer Grinder* destroyed 12 trucks in *Steel Tiger* East.

9 Nov: *The Exterminator* destroyed 19 trucks in *Steel Tiger* East.

19 Nov: Black Crow testing complete.

21 Nov: Two AC-123K "*Black Spot*" aircraft join 16 SOS.

5 Dec: *Thor* joined the 16 SOS for operational test.

7 Dec: Lt Col Berdell P. Brevig assumed command of the 16 SOS from Lt Col Thomas Simone.

14 Dec: *Ghost Rider* encountered first heavy AAA of dry season.

30 Dec: *Thor* kills 7 trucks and destroys a massive storage area.

1970

15 Jan: *Thor* sets new single mission truck kill record (30 destroyed, 2damaged).

26 Jan: *The Warlord* engaged by record AAA fire and suspected SA-2 launch.

31 Jan: *The Exterminator* and escorts destroyed 27 trucks of a 34 truck convoy.

14 Feb: *Thor* sets new single mission truck kill record (43 destroyed, 2 damaged).

18 Feb: *The First Lady* hit by 37mm AAA

19 Feb: *The Exterminator* and *Ghost Rider* were hit by 37mm AAA.

24 Feb: *The Warlord* located and destroyed camouflaged truck park and storage area

Apr: Maj R. P Hight assumed command of the 16 SOS.

22 Apr: *The Warlord* was shot down by 37mm AAA.

6 May: *Ghost Rider* battle damaged by AAA.

Jun: All aircraft sent to CONUS mod facilities for upgrade into *Pave Pronto* configuration

Jul: Lt Col Y. A. Tucker assumes command of the 16 SOS

21 Oct: *The First Lady* arrived on station at Ubon.

22 Oct: General Mayer, USAF Vice Chief of staff visited the 16 SOS.

22 Oct: *Thor* arrived on station at Ubon.

2 Nov: *Gomer Grinder* arrived on station at Ubon.

4 Nov: *Night Stalker* arrived on station at Ubon.

6 Nov: Secretary of the Air Force Dr. Robert Seamans Jr. visited the 16 SOS and was briefed by Lt Col Tucker.

7 Nov: *The Exterminator* arrived on station at Ubon.

7 Nov: General John D. Ryan Chief of Staff USAF visited the 16 SOS.

9 Nov: *Thor* was credited with destroying the first truck of the season for the 16 SOS.

16 Nov: Dr. John Kirk, MACV Science Advisor, visited the 16 SOS.

17 Nov: *Midnight Express* arrived on station at Ubon.

1 Dec: *Raid-Kills 'Em Dead* arrived on station at Ubon.

1 Dec, ASD assistance team arrived Ubon to solve fire-control issues. They remained until 17 Dec.

6 Dec, *Midnight Express* sighted, attacked and destroyed 12 trucks...the first big *Pave Pronto* mission.

12 Dec: *Raid-Kills 'Em Dead* received battle damage by 37mm AAA.

16 Dec: Thor destroyed 31 trucks and damaged two.

1971

5 Jan: *Grim Reaper* arrived at Ubon (flew first combat mission on 11 Jan).

11 Jan: Lt Gen Wells, USAF IG visited the 16 SOS.

14 Jan: *Proud Warrior* flew a record-breaking mission, destroying 57 enemy trucks and damaging 8. In addition, the escorts destroyed 7 and damaged 3 for a total of 75 tucks destroyed or damaged.

15 Jan: The Director of Operations for Gunships and Bombers directed a "hard crew" concept be established.

17 Jan: *Prometheus* arrived at Ubon (flew first combat mission on 20 Jan).

21 Jan: Gen Momyer, Commander TAC, visited the 16 SOS and received a briefing from the squadron commander, Lt Col Tucker.

21 Jan: *Prometheus* received battle damage in the leading edge of the right wing from 37mm shrapnel. Repairs were completed 22 Jan.

23 Jan: *Prometheus* again received battle damage. A 37mm projectile penetrated the right lower fuselage forward of the right main gear causing an 8–10 inch hole. Repair completed 2 Feb.

25 Jan: *Ghost Rider* received battle damage. A direct hit was taken in #1 engine by a 37mm round.

30 Jan: Lt Col Kenneth G. Harris assumed command of the 16 SOS.

2 Feb: Lt Gen Lavelle, Vice Commander PACAF visited the 16 SOS.

3 Feb: *Grim Reaper* and its escort, F4C, call sign Cash 4, successfully destroyed a 37mm gun using a laser guided bomb. This was the first time the AC-130 used this system under combat conditions.

16 Feb: *Proud Warrior* had a minor aircraft accident caused by a failed wheel rim while taxiing.

28 Feb: *Prometheus* received battle damage; the copilots window was shattered by 37mm shrapnel.

2 Mar: *Midnight Express* was fired on by two Surface to Air Missiles near Tchepone, Laos.

21 Mar: *Odin* arrived at Ubon (flew first combat mission on 27 Mar).

22 Mar: *Jaws of Death* arrived at Ubon (flew first combat mission on 26 Mar).

26 Mar: *The First Lady* received battle damage. A 37mm round entered under the cockpit and severed the utility hydraulic line. Minor difficulty was encountered lowering the main and nose gear. No casualties or injuries occurred.

30 Mar: *Proud Warrior* returned to Ubon after #3 40mm gun exploded. A small fire resulted with large holes in the left side of the fuselage and left inboard flaps. SSgt Adrian Richardson, gunner, was knocked unconscious. He was administered oxygen and first aid, and sustained only minor injuries.

Note: Data not available from April-September, 1971.

28 Jul: *Proud Warrior* received three .50 caliber hits in the right wing, presumably on final approach.

2 Oct: Contract Modification Team arrived on station to begin modification of 8 remaining AC-130A aircraft.

18 Oct: *Ghost Rider* arrived on station from CONUS modification program.

20 Oct: *Night Stalker* arrived on station from CONUS modification program.

22 Oct: Third AC-130A arrived on station from CONUS modification program.

26 Oct: *Thor* arrived on station configured with "GLINT" and "GREEN WEENIE".

25 Oct: The first AC-130E (*Excalibur*) and crew arrived on station.

1 Nov: *Excalibur* flew first combat mission.

6 Nov: Second AC-130E (*The Destroyer*) and crew arrived on station.

10 Nov: Third AC-130E (*Grave Digger*) and crew arrived on station.

12 Nov: *Orion The Hunter* received minor damage from suspected 50 caliber on final approach.

13 Nov: *Prometheus* returned to Ubon minus props on #3 & 4 engines.

19 Nov: The 16 SOS mascot "Ken" survived an attack by anonymous 8TFW fighter "jocks."

21 Nov: Fourth AC-130E (*Heavy Metal*) arrived on station.

25 Nov: AC-130E destroyed 9 and damaged 16 trucks in *Barrel Roll.*

27 Nov: *Raid-Kills 'Em Dead* received battle damage in fuselage and left wing from 37mm AAA while flying in VR-4 *Steel Tiger East.*

2 Dec: the 5th AC-130E (*Bad Company*) arrived on station.

10 Dec: Spectre 02 received over 1,400 rounds on AAA on a mission to *Barrel Roll.*

24 Dec: SSgt Ronald L. Hastings, Illuminator Operator, had his face shield knocked off by an AAA burst. He was not injured. Sgt James E. Smith, Illuminator Operator, was hospitalized for 2 days as a result of being thrown against the ramp during an evasive maneuver.

1972

1 Jan: Col Harry C. Canham, assumed command of the 16 SOS.

Feb 17: The first 105mm gun arrived at Ubon and was installed on *Iron Maiden.*

Mar 15: *Iron Maiden* received battle damage from 57mm AAA.

Mar 25: *Jaws of Death* was hit in the right wing, between the engines.

Mar 29: *Prometheus* was shot down by SA-2 Surface to Air Missile (SAM

Mar 30: *The Destroyer* was shot down by 57mm AAA.

Jun: Col J. E. Davis assumed command of the 16 SOS.

Jun 18: *Odin* was shot down by SA-7 missile near An Loc, Vietnam.

Note: Data not available for June-Sep, 1972.

Oct 1: *Excalibur*, killed 114 enemy during a TIC mission in Cambodia.

Oct 2: *Excalibur* assisted in base defense of Udorn AB using their 2KW light and flares.

Oct 11: *The Exterminator* had an SA-7 fired at it in the Saigon area.

Oct 17: *Ghost Rider* had a SA-7 fired at it in Cambodia.

Oct 20: *Proud Warrior* assisted in base defense of Korat AB using 2KW light.

Oct 26: *Ghost Rider* destroyed one truck and damaged five. It also destroyed a supply cache.

Oct 30: *Iron Maiden* destroyed two tanks and damaged one northeast of Pakse.

Nov 2: *Grave Digger* destroyed three trucks.

Dec 21: *Thor* was hit by 37mm AAA, then exploded and crashed while returning to base.

1973

Jan: The first AC-130H equipped with T56A-15 engines arrives at Ubon.

Jan 28: Vietnam cease fire declared.

Feb: Three AC-130H CROC arrive at Ubon.

Feb 22: Laos cease fire declared.

April 26: 16 SOS tasked to support *Operation Eagle Pull*, but the mission does not materialize.

Jul 6: 16 SOS achieved milestone of 1000 consecutive combat missions without an abort.

Aug 14: Aircraft 69-6574 flew the last combat mission of the war in Southeast Asia.

1974

April 19-28: Two AC-130H crews participated in a fire power demonstration for the President of Korea, Commander In Chief UN Command and their staffs.

July: 16 SOS moved from Ubon to Korat Royal Thai Air Base, Thailand.

The 16 SOS traveled to multiple locations in Southeast Asia to perform training and instruct other organizations of gunship capabilities.

1975

Apr 12: 16 SOS participated in *Operation Eagle Pull*, the evacuation of Phnom Pen, Cambodia.

Apr 29: 16 SOS participated in *Operation Frequent Wind*, the evacuation of Saigon.

May 13–15: 16 SOS participated in the rescue of the crew of the *SS Mayaguez*.

Jun 28: Last AC-130A departed Ubon for Duke Field, FL.

Jun 30: The 415th SOTS was deactivated at Hurlburt Field, FL.

Dec 9: Last AC-130H departed Ubon for Hurlburt Field, FL.

ACKNOWLEDGMENTS

The author is especially indebted to the following individuals who shared their recollections, expertise, and effort to ensure an accurate and truthful story was told. Most reviewed numerous manuscript drafts to ensure the highest level of comprehensive AC-130 history was captured.

A

Mr. William Alexander

B

Lt Col Charles Baertl, USAF (Ret.)
CMSgt Don Boudreaux, USAF (Ret.)
MSgt Dave Burns, USAF (Ret.)
Sgt Dennis Bivens, USAF
Mr. Ray Brandes

C

Col Gary Chandler, USAF, (Ret.)
Lt Col John Cahoon, USAF (Ret.)
Mr. Allen Cates, Air America
Lt Col Dave Coombs, USAF (Ret.)
Mr. Jeff Clark

D

Lt Col Richard Daugherty, USAF (Ret.)
MSgt Robert Davis, USAF (Ret.)

F

Lt Col Bernard "Bernie"
Frakes, USAF (Ret.)
Lt Col Larry Flinn, USAF (Ret.)
CMSgt Wayne Fisk, USAF (Ret.)
Sgt Ken Felty, USAF

G

CMSgt Mike Gilbert, USAF (Ret.)
Mr. Tom Guercio

H

SMSgt Frank Hanna, USAF, (Ret.)

J

Sgt Len Johnson, USAF

K

Col James Kyle, USAF, (Ret.)
Lt Col Richard "Dick"
Kauffman, USAF (Ret.)
Ms. Katrina Kiecker

L

CMSgt Bob Laymon, USAF (Ret.)

P

Maj Gen Robert Patterson, USAF (Ret.)

SMSgt William Patterson, USAF (Ret.)
SSgt Joe Percivalle, USAF
Sgt Larry Peterson, USAF
SFC Emil Praslick, USA (Ret.)
Mr. Merle Pribbenow, CIA (Ret.)

R

Lt Col Robert Reneau, USAF (Ret.)
Maj Hank Rother, USAF (Ret.)
1Lt Donald Raatz, USAF
SMSgt Ira Roseberry, USAF (Ret.)

S

Lt Col William Schwehm, USAF (Ret.)
Maj Gerald Stiles, USAF (Ret.)
Sgt Edward Sheridan, USAF
Mr. Dennis Santiago

T

Col Cal Taylor, USAF (Ret.)

W

Lt Gen Donny Wurster, USAF (Ret.)
LTC Mike Winslow, USA (Ret.)
MSgt Richard "Willy"
Williams, USAF (Ret.)
MSgt David Wilson, USAF, (Ret.)
Ms. Sarah Walter

Z

Lt Col Henry Zeybel, USAF (Ret.)

GLOSSARY

AAA:	Anti-Aircraft Artillery
ABCCC:	Airborne Battlefield Command and Control Center
ABGD:	Air Base Ground Defense
AC:	Aircraft Commander
ACC:	Air Component Commander
ADO:	Assistant Director of Operations
ADOC:	Air Defense Operations Center
ADM:	Admiral (rank)
ADVON:	Advance Echelon
ADSID:	Air Delivered Seismic Intrusion Detector
AFAL:	Air Force Avionics Laboratory
AFMPC:	Air Force Manning and Personnel Center
AFRES:	Air Force Reserve
AG:	Aerial Gunner/Airborne Gunner
AGL:	Above Ground Level
AIB:	Accident Investigation Board
ALE:	Air Launched Expendable
AP:	Armor Piercing
AOR:	Area Of Responsibility
APC:	Armored Personnel Carrier
AP-I:	Armor Piercing-Incendiary
ARVN:	Army of the Republic of Vietnam

<u>ARRS:</u> Aerospace Rescue and Recovery Squadron
<u>ASD:</u> Aeronautical Systems Division
<u>ATO:</u> Air Tasking Order
<u>AWACS:</u> Airborne Warning And Control System
<u>BC:</u> Black Crow (sensor)
<u>BDA:</u> Bomb Damage Assessment/
Battle Damage Assessment
<u>C2:</u> Command and Control
<u>C3:</u> Command Control and Communications
<u>CAPT:</u> Captain (rank)
<u>CAS:</u> Close Air Support
<u>CCTS:</u> Combat Crew Training Squadron
<u>CEP:</u> Circular Error Probable
<u>CIA:</u> Central Intelligence Agency
<u>CIDG:</u> Civilian Irregular Defense Group
<u>CINCPACAF:</u> Commander In Chief Pacific Air Forces
<u>COA:</u> Course Of Action
<u>CONUS:</u> Continental United States
<u>CP:</u> Copilot
<u>CQ:</u> Charge of Quarters
<u>CROC:</u> Combat Required Operating Capability
<u>CTZ:</u> Corps Tactical Zone
<u>CWO:</u> Chief Warrant Officer
<u>DET:</u> Detachment
<u>DMZ:</u> De-Militarized Zone
<u>DoD:</u> Department of Defense
<u>EW:</u> Early Warning/Electronic Warfare
<u>EWO:</u> Electronic Warfare Officer
<u>ECM:</u> Electronic Countermeasures
<u>FAC:</u> Forward Air Controller
<u>FCO:</u> Fire-control Officer

FDO: Fire Direction Officer
FSO: Fire Support Officer
FE: Flight Engineer
FO: Forward Observer
FPS: Feet Per Second
FRAG: Fragmentary Mission Execution Order
FSO: Fire Support Officer
G-Force: Gravitational Force
GLINT: Gated Laser Illuminator for Night Television
HE: High Explosive
HEAT: High Explosive Anti-Tank
HEP-T: High Explosive Plastic-Tracer
HE-P: High Explosive-Plugged (plugged tracer cavity)
HE-I: High Explosive-Incendiary
HE-I-P: High Explosive-Incendiary-Plugged (plugged tracer cavity)
HE-I-T: High Explosive-Incendiary-Tracer
HE-I-T-NSD: High Explosive-Incendiary-Tracer-Non Self Destroying
HUD: Head Up Display
HUMINT: Human Intelligence
IFE: In Flight Emergency
IFF: Identification Friend or Foe
IMC: Instrument Meteorological Conditions
IO: Illuminator Operator
IR: Infrared (sensor/operator)
ISC: Infiltration Surveillance Center
KBA: Killed By Air
KIA: Killed In Action
KIAS: Knots Indicated Air Speed
KW: Kilowatt

LAP: Load Assemble Pack
LAU: Launch Adapter Unit
LGB: Laser Guided Bomb
LLLTV: Low-light-level-television (sensor)
LM: Loadmaster
LNO: Liaison Officer
LP: Listening Post
LTD: Laser Target Designator
LTD/R: Laser Target Designator/Ranger
LTV: Ling-Temco-Vought
LWCP: Loader's Weapon Control Panel
LZ: Landing Zone
MIA: Missing In Action
MIL: Milliradians
MK: Mark
MM: Millimeter
MSL: Mean Seal Level
NASA: National Aeronautics and Space Administration
NAV: Navigator
NI-CAD: Nickel-Cadmium
NOD: Night Observation Device
NVG: Night Vision Goggles
OP: Observation Post
OPCON: Operational Control
OPLAN: Operation Plan
OPORD: Operations Order
OPSEC: Operational Security
PACAF: Pacific Air Forces
PAVN: People's Army Vietnam
PCS: Permanent Change of Station
PGU: Peculiar Gun Unit (ammunition for specific gun)

POL: Petroleum, Oils and Lubricants
POW: Prisoner Of War
RNO: Results Not Observed
ROE: Rules Of Engagement
RPG: Rocket Propelled Grenade[80]
RTB: Return To Base
RWR: Radar Warning Receiver
SAM: Surface to Air Missile
SAR: Search And Rescue
SECAF: Secretary of the Air Force
SF: Special Forces
SOF: Special Operations Forces
SOTS: Special Operations Training Squadron
SOW: Special Operations Wing
SPA: Surprise Package Aircraft
SPC: Specialist (Army)
SPO: Systems Program Office
SQUAWK: Select transponder code
TAC: Tactical Air Command
TACAN: Tactical Air Navigation
TACC: Tactical Air Control Center
TEMIG: Tactical Electromagnetic Impulse Generator
TFW: Tactical Fighter Wing
TG: Transportation Group
TIC: Troops-in-contact
TOT: Time On Target
TP: Target Practice
TP-T: Target Practice-Tracer

80 The Russian acronym RPG indicates *Ruchnoy Protivotankovy Granatomyot* (hand-held anti-tank grenade launcher), but the acronym is widely misidentified in the west as Rocket Propelled Grenade.

TS:	Top Secret
UTM:	Universal Traverse Mercator
USA:	United States Army
USAF:	United States Air Force
USAFSS:	United State Air Force Security Service
USMC:	United States Marine Corps
USN:	United States Navy
USO:	United Services Organizations
VAQ:	Visiting Airman Quarters
VDC:	Volts Direct Current
VHF:	Very High Frequency
VNAF:	Vietnamese Air Force
VR:	Visual Reconnaissance (area)
WP:	White Phosphorus

REFERENCES

The Development and Employment of Fixed-Wing Gunships, Jack S. Ballard

16th Special Operations Squadron Unit History Reports, 1968-1973

Bare Feet, Iron Will, James G. Zumwalt

Tactical Air Command Formal Flying and Related Training Publication, student text, USAF

Operational Training Course AC-130H Weapons Mechanics, Department of the Air Force

PAVN History Archives/ Mr. Merle L. Pribbenow

Project Checo Page Aegis Weapon System, July, 1973

Project Corona, Oral History of Lt Col Loyd J. King, 1973

Project Corona, Oral History of MSgt Arthur Humphrey, 1975

Official statement of Sgt Carl E. Stevens, 27 Dec, 1972 (declassified 31 Dec, 1980)

Col Harry Williamson, End of Tour Report, 1973

U.S., Air Force, History of the 388th Tactical Fighter Wing, 1 April 1975-30 June 1975, Vol II

The Mayaguez Story, Sea-Land Service, Inc. McEvoy

The true story of Operation Tailwind,
Maj (Ret.) John Plaster USAR

INDEX